KNOWLEDGE-BASED EXPERT SYSTEMS IN INDUSTRY

ELLIS HORWOOD BOOKS IN INFORMATION TECHNOLOGY
General Editor: Dr. JOHN M. M. PINKERTON, Principal, McLean Pinkerton Associates, Surrey, (formerly Manager of Strategic Requirements, International Computers Limited)
MODELLING HUMAN SPEECH COMPREHENSION: A Computational Approach
E. BRISCOE, Department of Linguistics and Modern English Language, University of Lancaster
PRACTICAL MACHINE TRANSLATION
D. CLARKE and U. MAGNUSSON-MURRAY, Department of Applied Computing and Mathematics, Cranfield Institute of Technology, Bedford
KNOWLEDGE-BASED EXPERT SYSTEMS IN INDUSTRY
J. KRIZ, Head of AI Group, Brown Boveri Research Center, Switzerland
BUILDING EXPERT SYSTEMS: Cognitive Emulation
P. SLATTER, Product Designer, Telecomputing plc, Oxford
SPEECH AND LANGUAGE-BASED COMMUNICATION WITH MACHINES
J. A. WATERWORTH, British Telecom Research Laboratories, Ipswich

KNOWLEDGE-BASED EXPERT SYSTEMS IN INDUSTRY

Editor:
JIRI KRIZ, M.Sc., Ph.D.
Head of Artificial Intelligence Group
Brown Boveri Research Center
Baden, Switzerland

ELLIS HORWOOD LIMITED
Publishers · Chichester

Halsted Press: a division of
JOHN WILEY & SONS
New York · Chichester · Brisbane · Toronto

First published in 1987 by
ELLIS HORWOOD LIMITED
Market Cross House, Cooper Street,
Chichester, West Sussex, PO19 1EB, England
The publisher's colophon is reproduced from James Gillison's drawing of the ancient Market Cross, Chichester.

Distributors:
Australia and New Zealand:
JACARANDA WILEY LIMITED
GPO Box 859, Brisbane, Queensland 4001, Australia
Canada:
JOHN WILEY & SONS CANADA LIMITED
22 Worcester Road, Rexdale, Ontario, Canada
Europe and Africa:
JOHN WILEY & SONS LIMITED
Baffins Lane, Chichester, West Sussex, England
North and South America and the rest of the world:
Halsted Press: a division of
JOHN WILEY & SONS
605 Third Avenue, New York, NY 10158, USA

© 1987 J. Kriz/Ellis Horwood Limited

British Library Cataloguing in Publication Data
Knowledge-based expert systems in industry. —
(Ellis Horwood books in information technology)
1. Industry — Data processing
2. Expert systems (Computer science)
I. Kriz, Jiri
338'.06   HD45.2

Library of Congress CIP available

ISBN 0–7458–0188–9 (Ellis Horwood Limited)
ISBN 0–470–20833–3 (Halsted Press)

Phototypeset in Times by Ellis Horwood Limited
Printed in Great Britain by R. J. Acford, Chichester

COPYRIGHT NOTICE
All Rights Reserved. No part of this publication may be reproduced, stored in a retrieval system, or transmitted, in any form or by any means, electronic, mechanical, photocopying, recording or otherwise, without the permission of Ellis Horwood Limited, Market Cross House, Cooper Street, Chichester, West Sussex, England.

Contents

Foreword 7

Preface 9

1. **Knowledge-based systems in industry: Introduction** 11
   J. KRIZ
   Brown Boveri Research Center, Baden, Switzerland

2. **The limitations of rule-based expert systems** 17
   J. L. ALTY
   Turing Institute and University of Strathclyde, Glasgow, United Kingdom

3. **A logic programming system for KBMSs—Educe** 23
   J. BOCCA
   European Computer-Industry Research Centre, München, Germany

4. **Extracting Parallelism from Sequential Prolog : Experiences with the Berkeley PLM** 36
   W. CITRIN,
   University of California, Berkeley, California, United States of America

5. **DEDALE: an expert system in VM/Prolog** 57
   Ph. DAGUE, Ph. DEVÈS,† Z. ZEIN, J. P. ADAM
   IBM France, Scientific Center, Paris; †Electronique Serge Dassault, Paris, France

6. **Logic-based tools for building expert and knowledge-based systems: Successes and failures of transferring the technology** . . . 69
   P. HAMMOND
   Imperial College, London, United Kingdom

7. **Model guided interpretation based on structurally related image primitives** 91
   G. MADERLECHNER†, E. EGELI‡, F. KLEIN‡
   †Siemens AG, München, Germany; ‡ETH Zürich, Switzerland

8. **Modula--Prolog: A programming environment for building knowledge-based systems** 98
C. MULLER
Brown Boveri Research Center, Baden, Switzerland

9. **CONAD: A knowledge-based configuration adviser, built using Nixdorf's expert system shell TWAICE** 111
S. E. SAVORY
Nixdorf Computer AG, Paderborn, Germany

10. **A Prolog frame system for knowledge-based design and diagnosis** . 117
H. SUGAYA
Brown Boveri Research Center, Baden, Switzerland

11. **Application of knowledge-based planning systems** 130
A. TATE
Artificial Intelligence Applications Institute, University of Edinburgh, Edinburgh, United Kingdom

12. **A prototype expert system for configuring technical systems** . . . 145
M. VITINS
Brown Boveri Research Center, Baden, Switzerland

Index 159

Foreword

Artificial intelligence has been the subject of basic research for more than 20 years. During the last 5 years, the practical applications and economic advantages of artificial intelligence approaches to computer programming and knowledge processing have come into view and triggered large research programs funded by governments and industry. At present, artificial intelligence is a booming research area, within which the sub-field of knowledge processing draws particularily great attention.

Knowledge-based systems are intelligent computer programs which are able to facilitate the mental work of the human being by providing decision support and consultation, based on conclusions drawn from stored knowledge about a certain domain. Flexible human–computer dialogs and problem-oriented languages for knowledge representation are supported in such a manner that these systems may be used and maintained by the domain expert not having an education in computer programming.

Recent research results have proven the feasibility of knowledge-based systems for engineering and business administration. It has become evident that knowledge-based systems offer a new quality of computer assistance, which can significantly facilitate the routine mental work of engineers and managers in tasks such as product design, project planning, manufacturing control, plant operation and fault diagnosis. However, we should be aware that the state-of-the-art in knowledge engineering is still a preliminary one and that most applications reported represent carefully restricted experiments. Further research combined with extensive pilot applications is necessary to yield large-scale advantages from the fascinating features of computerized knowledge processing.

Brown Boveri recognized the importance of knowledge-based engineering and planning tools several years ago. In 1984, a research program in knowledge-based systems, which later led to the establishment of the Artificial Intelligence Group, was launched at the Computer Science Department of the Brown Boveri Research Center. Since its conception, this research group has been continuously amplified and is presently one of the fastest growing groups within the Research Center.

We are very pleased that Brown Boveri could actively participate in the rapid progress of applied artificial intelligence and could organize this Workshop on Knowledge-Based Systems in Industry. The meeting was attended by 140 scientists and engineers from various countries. We would like to take this opportunity to express our sincere thanks to every Workshop participant. We are especially grateful to the authors for having spared no effort in preparing their papers. The high quality of their work is reflected in the contents of the present volume.

The organization of the workshop, the layout of the program and the contact with the speakers were the responsibility of Dr J. Kriz, leader of the Artificial Intelligence Group. His careful and competent preparation was instrumental in the success of the meeting and in the preparation of this volume. We also thank Miss D. Stadelmann for her smooth completion of the administrative portion of the meeting.

R. Güth
Head of Computer Science Department
Brown Boveri Research Center

A. P. Speiser
Director of Corporate Research
BBC Brown, Boveri & Co., Ltd.

Preface

1. WORKSHOP

The purpose of the Workshop held in Brown Boveri Research Center, Baden-Dättwil, Switzerland, on June 9, 1986 was to provide a forum for scientists and engineers to present and discuss industrial applications of Artificial Intelligence (AI), in particular knowledge-based systems. Invited speakers reported on the state of the art, current research activities and possible future trends. Emphasis was put on realizations of knowledge-based systems, and on practical experiences with AI techniques and tools. Topics from the following fields were addressed:

— knowledge-based systems for configuration, planning and diagnosis
— expert system shells and environments
— knowledge acquisition and knowledge representation techniques
— fast prototyping.

Since the Workshop was the first meeting of this kind in Switzerland many participants from Swiss industrial companies used this opportunity to get informed on the subject of knowledge-based systems. Many companies are faced with this new information processing technology and are evaluating its potential applications and economic benefits. For those who are entering this new field, the following brief introduction and survey, accompanied by a list of introductory literature, should facilitate their endeavour and comprehesion.

2. ACKNOWLEDGEMENTS

I was delighted that such well-known experts in their respective fields accepted the invitation and contributed to the success of the Workshop. I would like to thank the contributing authors for their great effort in preparing their papers.

I further thank the board of directors and the staff of the Brown Boveri Research Center for all the support which made the organization of the Workshop possible. In particular, I would like to thank Prof. Dr. A. P. Speiser, Director of Corporate Research, and Dr. R. W. Meier, Deputy Director of Corporate Research, for enabling and sponsoring the Workshop. I am very grateful to Dr. R. Güth, Head of the Computer Science Department, for providing a creative and highly motivating research atmosphere and for making the AI research in his department possible. I am very indebted to the members of the AI Group, D. Künzle, C. Muller, A. Repenning, Dr. H. Sugaya and Dr. M. Vitins, for many valuable suggestions and helpful support with the preparing and organizing of the Workshop. I thank Miss D. Stadelmann for her excellent administrative work. I acknowledge the professional assistance of the administrative and technical staff of the Brown Boveri Research Center.

Baden-Dättwil
June 9, 1986

J. Kriz

1
Knowledge-based systems in industry: Introduction

J. Kriz, Artificial Intelligence Group, Brown Boveri Research Center, Baden-Bättwil, Switzerland

A *knowledge-based system (KBS)* denotes a computer program that can store knowledge of a particular domain and use that knowledge to solve problems from this domain in an intelligent way [Hayes-Roth & Waterman & Lenat; Nau]. A KBS contains at least the following two components:

— the *knowledge-base*, in which the domain-specific knowledge is stored in form of facts and rules,
— the *inference procedure* ("inference engine") which operates on the knowledge-base, performs logical inferences and deduces new knowledge by applying rules to facts until the posed problem is solved.

A KBS should further provide the following features:

— explanation of its behaviour on request by the user,
— user-friendly dialog that takes the user and the kind of the application into account,
— application-oriented knowledge representation language,
— possibility to easily modify and extend the knowledge-base during the lifetime of the system.

Knowledge-based systems are also called *expert systems* [Duda & Shortliffe] if they refer to problem solving in those areas and at that level of performance that is usually achieved by human experts. A generic KBS with the above features but with an empty knowledge-base is also called an *expert system shell*. It is suitable for building specific expert systems by supplementing the particular knowledge base.

KBSs represent one of the application-oriented fields of artificial intelligence [Barr & Feigenbaum; Nilsson; Rich; Winston]. Although most KBSs, which have been realized so far, have a prototype character, there are several KBSs that have achieved a high maturity and are in practical use.

These systems are applied in fields like medical diagnosis [Shortliffe], chemical analysis [Lindsay & Buchanan & Feigenbaum & Lederberg], geological exploration [Duda & Gaschnik & Hart], configuration of computers [McDermott], monitoring and preventive maintenance of large technical equipments [Fox & Lowenfeld & Kleinosky].

KBSs can have a significant economic impact on the industry in the near future. It can be expected that still more fully operational KBSs will be realized, commercialized, and applied in the industrial production. There are several reasons for this. On one side, there has always been a great demand of acquiring, formalizing and utilizing the expert's know-how. This demand will increase due to the necessity of reducing production costs by rationalization. Also the tackling of still more complex tasks, which cannot be solved by human experts alone, require the support by expert systems. Furthermore, the conventional software engineering approach has often failed, because the problem areas are large, ill-structed and difficult to formalize in one step. On the other side, advances in AI, knowledge engineering, programming methodologies and computing resources have facilitated the realization of KBSs.

In the realization of KBSs the following AI-methods, techniques and tools are employed:

Symbolic processing
AI programming is concerned with manipulating not only numbers but general abstract quantities. Examples of typical objects are S-expressions (nested lists in LISP) and terms (nested functional symbols in Prolog). Examples of typical operations are list manipulation, matching and resolution. Symbolic processing is assumed to be a necessary and sufficient condition for a machine to perform a general intelligent action (Newell & Simon].

Declarative programming
The most important AI programming languages are LISP [McCarthy et al.; Winston] and Prolog [Colmerauer; Clocksin & Mellish]. Pure LISP is based on the concept of mathematical functions and realizes the concept of functional programming; Prolog is based on automatic theorem proving [Robinson] and logic programming [Kowalski 1974]. Both languages support declarative programming. This style describes the programming by specifying "what" should be solved rather than "how" it is to be solved. More precisely, in the declarative programming style the meaning of the objects is given by the static description of the objects themselves and is independent of other objects and of the program execution; there are no

side-effects. Hence, declarative programming supports the construction of modular and maintanable programs [Darlington; Clocksin].

Interactive programming environments
AI languages are primarily interpreted. The conventional execution cycle "edit→compile→link→run→debug" is hence shortened to "edit→run-→debug" that is performed without leaving and restarting the system. AI languages are usually embedded in interactive programming environments. Sophisticated environments facilitate and accelerate the programmer's work by providing syntax-oriented editors, debuggers, efficient interpreters and compilers, and advanced man-machine interfaces utilizing a high-resolution display and window techniques. Especially powerful environments have been realized on Lisp- and Prolog-machines that execute the respective programming languages very efficiently.

Incremental programming
The concepts of declarative and interactive programming support the incremental construction of programs. Programs can be easily modified and expanded. The idea of incremental programming is still further stressed in KBSs. Since the knowledge base is independent of the inference procedure, it can be realized stepwise during the development and simply updated during the lifetime of the system.

Fast prototyping
Declarative programming and easy program modifications allow the fast construction of functionally complete programs. These programs represent an executable problem specification and are well suited for testing the specification by simulation. When the possibly slow prototype becomes stable, it can be turned into an efficient program, for example by automatic compiling or even manual reprogramming.

Knowledge representation
A large amount of domain-specific knowledge is necessary for achieving high performance and intelligent behaviour of a KBS. Special methods and techniques [Nilsson] have been developed for representing knowledge, such as rules (productions) [Davis & King; Hayes-Roth], frames [Minsky; Fikes & Kehler] and logic [Kowalski 1979]. These methods allow the representation of the well-established and formalized knowledge about a particular area as well as the heuristic and fuzzy knowledge of domain experts.

Reasoning and problem solving
Problem solving can generally be described as a search in a state space [Rich]. A state of this space represents the facts known of a particular instant during the solution process. Transitions between states represent reasoning steps, i.e. the deduction of new facts by applying the rules of inference to the old facts. Symbolic processing is well suited for the implementation of the abstract search and hence for the realization of reasoning and logical inferencing.

Knowledge engineering environments and expert system shells
Based on the above AI-concepts and techniques, higher-level systems for knowledge engineering and expert systems have been designed and implemented in commercially available products. Such environments and shells usually include various paradigms of knowledge representation and reasoning.

Although the applied AI and KBSs have achieved a high standard, there are still many open problems that are the topic of current research. One of the more practical problems is the enormous demand of AI programs on the computing speed and memory. This problem is successfully approached along several lines:

— the originally purely declarative AI languages are enhanced with procedural features and even mixed with conventional languages in one integrated system,
— AI programs, which can be slow when interpreted, can automatically be transformed or compiled into fast procedural programs,
— conventional processors and memories become cheaper and more powerful, such that AI-languages and KBSs can even been implemented on personal computers,
— new computer architectures are being developed that should be more suited to symbolic and logic processing [Feigenbaum & McCorduck].

Other problems, whose solution cannot be expected early, concern the knowledge acquisition, representation of common sense knowledge, the utilization of "deep" knowledge for the reasoning from first principles, learning, vision and natural language understanding.

Applied AI and KBSs represent a promising approach for treating difficult, ill-structured problems whose solution requires domain knowledge, significant expertise, much experience and heuristic rules. KBSs have a high rationalization potential and therefore attract the interest of industry. Expectations of KBSs are high, still rising and can become unrealistic due to the great appeal of the topic. AI can do much but not everything — in particular, it cannot yield the general problem solver. Also, AI should not be viewed as the only and exclusive alternative but as an additional and complementary approach to conventional programming. If AI research and applications are treated realistically and seriously, the great potential of AI can best be utilized and appreciated.

REFERENCES

Barr, A., E. A. Feigenbaum (1981) *The Handbook of Arificial Intelligence,* Vols. 1–3, Pitman.

Clocksin, W. F., Mellish, C. S. (1984) *Programming in Prolog,* 2nd ed., Springer, Berlin.

Clocksin, W. F. (1984) Logic Programming and Prolog, in Chambers F. B. *et al.* (eds.) *Distributed Computing,* Academic Press, London, 79–109.

Colmerauer, A., Kanoui, H., Roussel, P., Pasero, R. (1973) Un System de Communication Homme-Machine en Francais, Groupe de Recherche en Intelligence Artificielle, Universite d'Aix-Marseille.

Darlington, J. (1984) Functional Programming, in Chambers, F. B. *et al.* (eds.) *Distributed Computing,* Academic Press, London, 57–77.

Davis, R., King, J. (1977) An Overview of Production Systems, in Elcock, E., Michie, D. (eds.) *Machine Intelligence 8,* Ellis Horwood, Chichester, 300–332.

Duda, R. O., Gaschnik, J. G., Hart, P. E. (1979) Model Design in the PROSPECTOR Consultant System, in Michie D. (ed.) *Expert Systems in Microelectronic Age,* Edinburgh University Press, 153–168.

Duda, R. O., Shortliffe, E. H. (1983) Expert Systems Research, *Science,* Vol. 220, No. 4594, 261–268.

Feigenbaum, E. A. and P. McCorduck (1983). *The Fifth Generation — Artificial Intelligence and Japan's Computer Challenge to the World,* Addison-Wesley, Reading, Mass.

Fikes, R., Kehler, T. (1985) The Role of Frame-based Representation in Reasoning, *Comm. ACM,* Vol. 28, No. 9, 904–920.

Fox, M. S., Lowenfeld, S., Kleinosky, P. (1983) Techniques for Sensor-Based Diagnosis, Proc. 8th Int. Joint Conf. Artificial Intelligence, Karlsruhe, 158–163.

Hayes-Roth, F., Waterman, D. A., Lenat, D. B. (1983) *Building Expert Systems,* Addison-Wesley, Reading, Mass.

Hayes-Roth, F. (1985) Rule-based Systems, *Comm. ACM,* Vol. 28, No. 9, 921–932.

Kowalski, R. A. (1974) Predicate Logic as Programming Language, Proc. IFIP 74, North-Holland, 569–574.

Kowalski, R. A. (1979) *Logic for Problem Solving,* North-Holland, Amsterdam.

Lindsay, R., Buchanan, B. G., Feigenbaum, E. A., Lederberg, J. (1980) *Applications of Artificial Intelligence for Chemical Inference: The DENDRAL Project,* McGraw-Hill, New York.

McCarthy, J., Abrahams, P. W., Edwards, D. J., Hart, T. P., Levin, M. I. (1962) LISP 1.5 *Programmer's Manual,* MIT Press, Cambridge, Massachusetts.

McDermott, J. (1982) R1: A Rule-Based Configurer of Computer Systems, *Artificial Intelligence,* Vol. 19, No. 1.

Minsky, M. (1975) A Framework for Representing Knowledge, in Winston, P. H. (ed.) *The Psychology of Computer Vision,* McGraw-Hill, New York.

Nau, D. S. (1983) Expert Computer Systems, *Computer,* Vol. 16, No. 2, 63–85.

Newell, A., Simon, H. A. (1976) Computer Science as Empirical Inquiry: Symbols and Search, *Comm. ACM,* Vol. 19, No. 3, 113–126.

Nilsson, N. J. (1982) *Principles of Aritificial Intelligence,* Springer, Berlin.

Rich, E. (1983) *Artificial Intelligence,* McGraw-Hill, New York.

Robinson, J. A. (1965) A Machine-oriented Logic Based on the Resolution Principle, *J. ACM,* Vol. 12, 23–41.

Shortliffe, E. H. (1976) *Computer-Based Medical Consultations: MYCIN,* Elsevier, New York.

Winston, P. H., Horn, B. K. P. (1984) *LISP,* 2nd ed., Addison-Wesley, Reading, Mass.

Winston, P. H. (1984) *Artificial Intelligence,* 2nd ed., Addison-Wesley, Reading, Mass.

2

The limitations of rule based expert systems

J. L. Alty, Turing Institute and University of Strathclyde, Glasgow, United Kingdom.

Expert Systems are currently the subject of an intense research effort and appear to be an area of applied Artificial Intelligence which may prove to be very cost effective. A recent survey by Pactel revealed that 10% of all large companies were currently experimenting with expert sytems techniques, that 30% more were expected to move into the area within the next 3 years, and that the majority of the remainder would seriously examine their potential within 5 years. The survey predicted a huge market for such systems by 1990. The Expert System approach has been extensively discussed (Alty and Coombs, 1984), (Waterman, 1986), (Weiss and Kulikowski, 1984), (Hayes-Roth, Waterman and Lenat, 1984), and need not be expanded here.

In a climate of such predictions and euphoria by Expert Systems designers it is appropriate to stand back a little and remind ourselves of the limitations of current techniques and examine how they might be overcome. Artificial Intelligence workers (or at least those who promote the application of their techniques) must beware lest the expectations of users are raised too high without proper justification.

THE RULE-BASED APPROACH AND EXPLANATION

Most current Expert Systems are based upon the Production Systems architecture (see Waterman (1986) for a discussion of architectures). Basically a set of production rules (or IF.. THEN rules) act upon a data-base of facts and relationships under some control regime to move a system from some start state to a given goal state. Such systems have the advantage of being able to operate either in a forward-chained mode (data driven) or backward-chained mode (goal driven) manner and are therefore able to solve a wide range of problems. This problem solving capability together with the explicit representation of the domain knowledge provides Expert Systems with many of their claimed benefits. In some systems the two reasoning directions may be intermixed as for example in the 'blackboard' approach (Erman 1980).

The above architectural features overcome a number of traditional problems. For example the traditional programmer might better be described as a knowledge 'butcher' rather than a knowledge representer. Much domain knowledge is discarded and, what is worse, a great deal of extraneous knowledge (e.g. computational knowledge) is added during the programming process. It is this latter knowledge which often makes program maintenance difficult. The explicit nature of the knowledge in Expert Systems also enables them to provide a form of 'explanation' or 'justification' on demand to the user. At any time a user may ask WHY or HOW? and a trace of the rules used to reach the current point in the consultation is displayed. Considerable claims are made about the usefulness of the explanation facilities provided by Expert Systems. It is therefore worthwhile to examine the claim in some detail because there are difficulties in the approach which are not often stressed. Regurgitation of a set of rules used to follow a particular reasoning path provide a very limited form of explanation. Firstly, the rules displayed are in the form in which the original expert placed them in the system. They may be meaningful to another expert but how meaningful might they be to a novice user?. In real-life experts do not explain or justify themselves in this way. They use an analogical approach referring to some well known concepts in the user's domain of experience. Unfortunately Artificial Intelligence has not yet come up with any real answers to the problem of providing a form of analogical reasoning.

The current explanation facilities therefore work well when the 'users' of the system understand the language used by the experts to describe their expertise. Such a situation is not uncommon — all doctors basically use the same medical language and approach so that an explanation given by an expert Dermatologist will be understandable to most non-specialist doctors. The medical format will be the same even if the content is new. This is one reason why the explanation facilities in systems like MYCIN worked well. However the explanation will not be very meaningful when expert and user talk 'different' languages.

In fact the dialogues provided by Expert Systems tend to be very rigid. Often the user is constrained and led by the dialogue. Even in systems where the user can "interrupt" to volunteer data the interruption is treated as a temporary diversion from the main line of the consulation. In field applications this rigid approach can cause a number of problems (Kidd and Cooper 1983). The approach is not well suited for supporting a number of users with different problem solving styles using the same knowledge base. Another consequence of the rigid approach inherent in most Expert Systems dialogues is the lack of support for a "consultative" interaction. Often users consult experts in order to receive assistance in solving their own problems yet this aspect of problem solving is neglected in current systems (Kidd, 1985).

EXPLICIT REPRESENTATION OF DOMAIN KNOWLEDGE

Another difficulty with current systems is the degree to which they actually store the domain knowledge in explicit form. Although most expert system designers recommend that domain dependent control structures should be avoided, such advice is not always followed. The MYCIN system illustrates this point. The domain rules are ordered in such a way that the order of the rules determines which rule in the conflict resolution set is actually fired. Thus the crucial knowledge in the system — what to do next — is implicit rather than explicit. If a user asks "why did you fire that particular rule?" the system can only answer "because it was there"!. Attempts have been made to overcome this difficulty by the use of meta-rules. These are rules about rules and the concept is exemplified in the NEOMYCIN system developed by Clancy (Clancy 1981).

The NEOMYCIN system attempts to make the strategic knowledge explicit by building a set of meta-rules above the MYCIN domain rules. It provides an explicit separation between the diagnostic strategy and the disease knowledge. Thus user can ask why a particular strategy was chosen and obtain WHY or HOW explanations in a similar manner to that of MYCIN for domain knowledge. A similar approach was used in GUIDON (an extension of MYCIN to provide tutorial instruction to medical students). Additional meta-rules were added to guide the system's tutoring and explanation mechanisms (Clancy 1979).

LIMITATION OF THE EMPIRICAL APPROACH

So many Expert Systems have now been produced using empirical associations involving sets of IF..THEN rules that the reader might overlook the fact that such a representation is only useful within a limited set of application domains. The production rule format is ideal for representing "ad-hoc" knowledge or rules of thumb. However the knowledge of any experts is not in this form. In many diagnostic domains, for example, the expert has a model of the system under investigation and his problem solving techniques involve working backwards with the model in order to isolate faulty components. In other words experts often use causal models in their reasoning processes.

Using a set of IF..THEN rules to describe an electric circuit would be extremely tedious. Experts do not store 'scientific' knowledge in this form, rather they know about Kirchhoff's Laws or Ohm's Law, or at least they have a good approximate model of what is going on. Likewise, an expert car mechanic does not only utilise a set of such rules to diagnose car faults. Such rules do exist but, in addition, the mechanic has a model of what is going on and uses this model to guide the diagnostic process. Furthermore, knowledge stored in production rules normally has little generality. It works well in the domain of its applicability but is not easily transferred into other domains. Model knowledge on the other hand can be partially transferred. A technician who is expert at solving problems on one particular circuit can usually use this knowledge to solve problems on other circuits as well.

A MODEL-BASED APPROACH

Recently research in Expert Systems has started to investigate approaches based upon reasoning from first principles, or at least using plausible models. Examples of this approach include SOPHIE (Brown, Burton and

DeKleer 1982) and CASNET (Kulikowski and Weiss 1982) was an early example.

The empirical association approach connects a set of symptoms with diseases (or causes). The IF part of the rule has been found from experience to result in the THEN part. The rule connects the two but does not embody the reason behind the connection. For many tasks a better approach is to reason from behaviour to structure or rather from misbehaviour to structural defect (Davis, 1984). This requires a language to describe function, one to describe structure and a set of principles for fault diagnosis. Davies (Davies 1984) has given an excellent account of such an approach which involves reasoning from first principles about electronic devices. A functional and physical description of the circuit is built up from basic units of description called modules. Modules at any level can have sub-structures so that models at different levels of abstraction are possible. Sets of production rules describe possible computations and inference which can be made about the circuit. For example (from Davis, 1984) an adder with two inputs (input-1, input-2) and an output (sum) can be described by three rules:

to get sum from (input-1, input-2) do (+input-1, input-2)
to get input-1 from (sum, input-2) do (−sum, input-2)
to get input-2 from (sum, input-1) do (−sum, input-1)

The first rule describes the actions of the circuit. The second and third rules are inference rules. No adder operates in this way but a designer uses such rules to work backwards from an output to an input.

Such sets of rules model the execution of the circuit and the inference mechanisms involved in trouble-shooting but there are difficulties. Faults on the circuit can be due to short-circuits at junctions or between connections. These are not in the original model. Since such errors can appear in many places they must be modelled in some way, yet if all the possible short-circuit conditions are required (through additional connections) the model becomes impossibly complex. Thus a model which contains all possible short-circuits is simply not possible. The problem is solved by tracing paths of causality rather than connection paths. A set of interaction models is postulated and some metric is used to order them. The lowest model of interaction is first used to endeavour to account for the observed behaviour (i.e. localised failure). Only if this fails is the next model used (in this case power failure). If this fails a model of interaction involving multiple errors is invoked. The next two-higher levels are concerned with assembly errors and design errors.

These various interaction models almost certainly mimic the actions of a good engineer. He does not approach a circuit problem by first assuming a design error. Simplifying assumptions are made about the nature of the fault. Only when this approach fails are these discarded and additional models of failure considered.

The above approach is certainly more general than the empirical rule-based approach. The problem solving strategies are valid over a range of devices and therefore aid carry-over of trouble-shooting knowledge into new situations. Furthermore traditional systems are difficult to construct. The symptom-cause rules have to be extracted from the expert on a case by case basis and are particularised to a particular device. The approach is therefore more likely to provide the basis for the development of more powerful diagnostic reasoning system.

CASNET (Weiss and Kulikowski, 1982) was developed to assist in the treatment of the eye disease Glaucoma. The knowledge is stored as a series of causal pathways in a plane of pathophysiological states. A causal pathway traces the progress of the disease as symtoms develop and two further planes of knowledge containing disease categories and symptoms are linked into this plane. The central plane contains a causal model of the progress of the disease. As additional evidence is gathered the actual disease category becomes clear. One problem with a disease like glaucoma is the slow progress of the disease over time. This model based approach is ideal for handling such a situation. For a simple explanation of the mechanisms used in CASNET see Alty and Coombs (1984).

Model-based reasoning poses problems for the traditional reasoning strategies of Expert Systems. Usually the reasoning is monotonic. Facts are instantiated and are then assumed to be true for the remainder of the reasoning process. In model based reasoning the different models represent different views of the system and may well be inconsistent. Thus facts established using one model may not be true according to some other model view. Human beings certainly appear to reason in this way. In model based reasoning we need to be clear what we can carry across from one model to another and what we cannot.

CONCLUSION

This chapter has attempted to highlight some current problem areas in Expert Systems design. Current technology has been very successful in solving problems in some very restricted domains. The problems mentioned above are now being addressed by workers in the fields. Their resolution will considerably extended the usefulness of the Expert System's approach.

REFERENCES

Alty, J. L. and Coombs, M. J. (1984); *Expert Systems, Concepts and Examples,* NCC Publications, The National Computing Centre, Oxford Road, Manchester, United Kingdom, pp. 209.

Brown, J. S., Burton, R. and Dekker, J. (1982); Pedagogical and knowledge engineering techniques in SOPHIE I, II and III, in Sleeman D. H. and Brown, J. S. (eds.) *Intelligent Tutoring Systems.* London, Academic Press.

Clancey, W. J. (1979); Tutoring rules for guiding a case method dialogue, *Int. J. Man Mach Studies,* Vol. II, pp. 25–49.

Clancey, W. J. and Letsinger, R. (1981); NEOMYCIN: Reconfiguring a rule-based expert system for application to teaching, IJCAI-7, pp. 829–836.

Davies, R. (1984); Reasoning from first principles in electronic trouble shooting, in *Developments in Expert Systems* (ed. Coombs, M. J.) London, Academic Press, pp. 1–21.

Erman, L. D. (1980); The HEARSAY-II speech-understanding system: integrating knowledge to resolve uncertainties, *Computing Surveys,* Vol. 12, No. 2, pp. 213–253.

Hayes-Roth, F., Waterman, D. and Lenat, D. (1983); *Building Expert Systems,* Addison-Wesley, Reading, Mass.

Kidd, A. L. (1985); The consultative role of an expert system, in Proc. Conf. BCS Human Computer Interaction Specialist Group: *People and Computers, designing the interface* (eds. Johnson, P. and Cook, S.) Cambridge University Press, London, pp. 246–263.

Kidd, A. L. and Cooper, M. B. (1983); Man-machine Interface for an Expert System, Proc. BCS Expert Systems 1983 Conference, Cambridge.

Kulikowski, C. A. and Weiss, S. (1982); Representation of expert knowledge for consultation: the CASNET and EXPERT projects, in Szolovits, P. (ed.) *Artificial Intelligence in Medicine,* Boulder, Colorado, Westview Press, pp. 21–55.

Waterman, D. (1985); *A Guide to Expert Systems,* Addison-Wesley, pp. 419.

Weiss, S. and Kulikowski, C. (1984); *A practical Guide to Designing Expert Systems,* New Jersey, Rowman and Allanheld.

Weiss, S. and Kulikowski, C. (1981); Expert consultantion systems: the EXPERT and CASNET projects, in Bond, A. H. (ed.) *Machine Intelligence,* Infotech State of the Art Reports, Series 9, No. 3, Pergamon Infotech Ltd., pp. 339–353.

3

A logic programming system for KBMSs — Educe

Jorge Bocca, European Computer-Industry Research Centre GmbH (ECRC), Arabellastr. 17, D-8000 Muenchen 81, West Germany.

1. INTRODUCTION

The need for an efficient logic programming data base (DB) system has been expressed by a number of researchers in the recent past [19, 14, 6, 4]. Given our objective of constructing a fully fledged data based management system (DBMS) based knowledge based management system (KBMS) at ECRC, this need became of particular importance to us. Hence Educe was designed and implemented.

According to the way they are implemented, logic programming DB systems can be divided into three categories [12]: systems that are constructed by the integration of a file system into a rule manager [15, 10], systems constructed by extended a DBMS to support rules and inference [12], and systems based on the full integration of the DBMS and the rule manager [1]. This last case is normally considered as the best possible solution of the three cases [12, 14], but unfortunately, also the hardest to implement. Educe belongs to this class of system.

The main requirements in the design of Educe appeared at two distinct levels, the physical and the logical levels. At the physical level, we were concerned with questions of efficiency, storage methods, optimization techniques, etc; while at the logical level, we were concerned with the way that users perceived the system, i.e. ways of querying and generally interacting with Educe's knowledge/data bases.

In designing and implementing Educe, we wanted to investigate and compare the basic techniques available to us at the physical and logical levels. Thus, at the physical level we implemented and have consequently evaluated a coupling and integration of Prolog and the external data base (EDB) [3, 7]; while at the logical level, we provided users of Educe with two contrasting data manipulation languages (DMLs): a loose and a close DML. These concepts and others, the basic techniques and the evaluation of them are the subject of this paper. Other papers which cover related work in this area and which are not directly discussed here, are [11, 20, 14, 19, 13, 8, 9, 16]. This paper is divided into five sections. The first section is this introduction; the second section describes Educe's design; the third section reviews our solutions to the different problems posed by the design of Educe; the fourth section describes those problems not yet solved and in fact, often unearthed by our use of Educe; and finally, section five presents conclusions and an outline of future work.

2. A DESCRIPTION OF EDUCE

Although the material in this section has been presented in more detail in [3, 4, 5], it has been included here for the purpose of making this paper self-complete.

2.1 The levels

When discussing the physical level in a logic programming system with Educe like characteristics, two components can be identified, the deductive part and the external data base (EDB). Depending on the way in which these components physically interact with each other, one can distinguish two extreme cases: coupling and integration. In a coupling the two parts can clearly be recognized as units in their own right, and for the purposes of the logic DB system as a whole, they collaborate together. In an integrated logic system by contrast, the deductive and the EDB components merge together and so become one unit.

In the case of Educe, we explored both forms of physical interaction between the deductive and the EDB components. For the deductive component we adopted and expanded the logic programming language Prolog, and for the EDB component we chose the relational data base management system Ingres. The choice of Prolog as the deductive component of Educe was based on its importance as a logic programming language. The selection of Ingres for the EDB component of Educe was more arbitrary. Without substracting from the merits of Ingres, we felt that any other relational data base management system (DBMS) would have been appropriate, provided that it had been implemented in a reasonably modular manner and used efficient data access methods.

At the logical level, we also identified two cases. Based on the degree of similarity to Prolog, we distinguish two data manipulation languages (DMLs): (i) a DML that follows the notation and conventions used for Prolog's facts, and (ii) a DML with syntax and conventions which are similar to the ones in DMLs for relational DBMSs. We refer to these two cases, as *close* DML and *loose* DML, respectively.

Perhaps the most obvious advantage of the close DML to the Prolog programmer is the high degree of transparency at the linguistic level. Normally, Prolog's rules and facts can be naturally mixed without explicit references to the DBMS.

Because of transparency, the close language appeared as a very attractive possibility to us. In particular, it was possible to implement in the first instance prototype systems in Prolog, without a DBMS facility [5, 17, 18]. Unfortunately, the advantages of the close DML were not exempt from some serious drawbacks. For instance, control predicates in Prolog restrict the applicability of optimization techniques based on query transformation [4]. Because of this loose coupling became an interesting possibility. We implemented the loose DML in Educe by the embedding of a high level non-procedural DML into Prolog. In this manner, we made the **relation** the atomic unit of access between control statements, so avoiding the problems created by the use of control predicates in the close DML. The embedded DML is a means of communication between Prolog and the EDB, with well defined entry and exit points.

Another important characteristic of the loose DML is that Prolog's variables are used in the specification of projections, and they can also be used in the specification of retrieval conditions (selection/join). The use of variables in the condition part of *retrieve* is essential in recursive definitions. In this manner, the use of Prolog's control predicates is forced beyond the boundaries of retrievals.

In designing Educe, we felt that there were no definitive answers to the dichotomies of coupling vs. integration and loose vs. close. However, we identified clear affinities between coupling and a loose DML, and also between integration and a close DML. Because of this, we implemented the first version of Educe along these lines, i.e. for the loose DML we used a coupling and for the close DML we used integration. At a later stage, in order to conduct some experiments on performance, we mapped the loose DML into close form and vice versa. This effectively allowed us to compare the evaluation strategy of *set retrievals* commonly used by DBMSs against the *one-tuple-at-a-time* evaluation strategy of Prolog. Such a comparison is needed for the development of efficient evaluation techniques for recursive queries. Recursion in the context of knowledge bases lends itself to many optimization techniques which are unknown in conventional DBMSs [17, 2].

2.2 The implementation

From the point of view of the implementation, an obvious method suggested itself for the loose DML. Provided that recursion is not allowed, a simple way to construct a loosely coupled system is by setting up two processes, one for the deductive component and one for the EDB component. These two processes exchange messages, i.e. queries and replies, through a channel of communication. This set up was accomplished in Educe by setting up one process for Prolog, as the deductive component, and one process for Ingres as the EDB component. Communication between the Prolog and the Ingres processes is by means of two pipes, one for queries and one for replies. To implement close integration in Educe, we decided to integrate the deductive component and the EDB component. For this, we detached the *access methods* module from the DBMS and attached it to Prolog. Then, in order to integrate loose coupling and close integration into one system, we considered the whole system as two concurrent processes, each running a DBMS on a common data base.

For the final architecture, we merged the three previous configurations. In the double DBMS configuration, we replaced one of the occurrences of the DBMS by our Prolog+AM configuration. This was possible because the Access Methods (AM) module of the Prolog+AM and the DBMS are identical replicas. In other words, Educe is constructed as two different but concurrent DBMSs with shared access to a common data base.

2.3 The evaluation strategy

Recursive definitions which include expressions in loose form are evaluated by a hybrid strategy. An evaluator was implemented for this purpose. The evaluator uses loose coupling for the non-recursive part of the definition, and then, for the recursive part, it uses the route provided by close integration for retrievals from the intermediate results.

More recently, the general stategy of evaluation of Educe has been reviewed. A module which maps expressions from one of the languages (loose or close) into the other has been built. This module allows Educe to select a route, either coupling or integration, entirely on the basis of expected performance.

3. PROBLEMS AND THEIR SOLUTIONS

As already mentioned in the previous section, the use of control predicates, such as cut in Prolog, causes serious problems to the optimization of queries. Because of this, we provided users with a loose DML so that interactions with the EDB appear boxed into logical units. In this scheme, optimization by query transformation is left to the underlying DBMS. However, because of the preferred usage of the close DML by Prolog programmers, the question of goal re-ordering in the close DML had to be reviewed [18]. Ideally, we would have liked to be able to transform a goal such as

 ?-employee (Name, Dept), ...,
    otherpred (..., Dept, ..),
    ..., Dept=production.

into the "*equivalent*" form

 ?-Dept=production, employee (Name, Dept),
    otherpred (..., Dept, ..), ...

so that the early instantiation of *Dept* could be used on an indexed search of those *employees* in the *production* department. This sort of transformation would allow us to make full use of the data structures and the access methods used by Educe. Unfortunately, because of the likely occurrence of control predicates in *otherpred*, the suggested transformation is not possible. However, we thought that it was still possible to make an effective use of Educe's access methods (AM). The obvious case is the one where the arguments are grounded on an indexed attribute. For example, to evaluate

 ?-employee (Name, production).

Educe can make use of an index on the *Dept* attribute. In order to take full advantage of Educe's AM, we stretched the previous idea to deal with range queries in the close DML. For this, we extended the Prolog syntax to accommodate these cases. Thus, expression like

?- earns(Name, Salary), Salary>1000.

should be written as

?-earns (Name, Salary>1000).

in order to take advantage of the performance benefits of an indexed search. However, the former expression is still valid, although expensive to evaluate.

Another apparently trivial problem is the correspondence of data types in Prolog and the underlying DBMS. In Educe we chose to map

DBMS	Prolog
character string	atom
integer	integer
real	real

Unfortunately, regardless of the particular components (DBMS and Prolog interpreter) the match is not always perfect. A common problem is that while the Prolog interpreter/compiler uses tags in the internal representation of values and their data types, this same information is normally kept by the schema maintenance mechanism of the DBMS. The consequence of this is that although the DBMS and the Prolog interpreter might use one word (in the machine) to represent integers, because of tags in the Prolog interpreter fewer bits are available to store values. The same holds true for *string-atom* and *real-real*, although with slight variations. This might not seem a serious problem for completely new applications, but it is a rather serious problem when a new application attempts to use an existing data base.

The question of data types leads into one more problem. Since we decided to store the intensional part as well as the extension of relations in the EDB, we needed a way of storing rules in the EDB. This sort of structured data types is not normally supported by a relational DBMS. In Educe, we solved the problem by creating a relation (*rulerel*) to store rules. The rules are stored as character strings and two new built-in predicates are used to map the rules into character strings (*swrite*) and vice versa (*sread*). Rules stored in the EDB are automatically used if no matching definition is found in main memory during evaluation of a goal. Extensions of relations kept in the EDB can be made transparent to Prolog users by defining a rule in *rulerel*.

The use of a relation to store rules in the EDB brings to the fore the incompatibility of the procedural evaluation of goals in Prolog with the concept of a relation as an unordered set. While the order of unit clauses would only alter the order in which answers are produced, the order in which general program clauses are chosen for evaluation might produce completely different answers, if any at all. For instance, choosing the recursive rule first for the evaluation of the classical *ancestor*, definition, could lead to an infinite loop, while choosing the non-recursive rule first would produce the correct answer(s). To get around this problem, users of Educe have to specify the order in which rules stored in the EDB should be used during evaluation of queries.

Still on the subject of rules in the EDB, two additional problems. As already mentioned, once rules are stored in the EDB they can be used like any other rule. A simple way of achieving this is to modify the top level of the Prolog interpreter, so that whenever no matching definition for a goal G can be found in main memory, the following program is executed:

```
interpret(G) :-        /*evaluate G*/
    functor(G, Label, N),
    /*find rule in EDB*/
    rulerel(Label, N, _, RuleinChars),
    /*convert string into rule*/
    sread(RuleinChars, Rule),
    (Head:- Body)=Rule,
    G=Head,
    call(Body).
```

Imagine that, among the several clauses defining **G**, at least one is recursive. The first problem is one of efficiency. In order to evaluate **G**, it is necessary to retrieve the recursive clause several times. The second problem is even more serious in a multi-user environment. Consider the case where one user is evaluating **G**, while a second user is modifying the recursive clause. This leads to a situation where, during the evaluation of **G**, two different recursive clauses are used. This evaluation of **G** might thus produce seriously wrong results. In order to solve these two problems, we decided to pre-load all the necessary definitions into main memory before starting the evaluation of the top goal. On return to the top level interpreter, and when no more answers are required-or no more answers can be found, the pre-loaded definitions can be erased from main memory. This scheme effectively freezes definitions until the evaluation of the goal(s) at the top level is completed. In addition, it solves the efficiency problem, since now only one access to each required clause in the EDB is made.

However, the above scheme is not enough. The above scheme would fail to evaluate correctly queries like

?-nrule(1, 'p(X) :- q(X).   '), p(Z).

even if the internal/external DB contained

q(a).
q(b).
:
:

The reason for the failure is due to the attempt to pre-load the definition for **p**, before the new rule (*nrule(..)*) is added to the EDB. Our final solution is to incrementally load the definitions from the EDB as they are required. This solves both of the above problems, efficiency and concurrent access, and at the same time it maintains the order of evaluation prescribed by Prolog.

A number of other techniques had to be devised in order to improve performance. In addition to the optimization techniques of the DBMS, Educe uses its own techniques. Particular attention was given to the recursive case, since this is an area outside the scope of conventional DBMSs. Four significant cases are discussed here.

The first case arises in queries involving one base relation and the boolean condition *true*. Although this case seems trivial and unlikely to be presented to Educe by users, it often arises as an intermediate step in the evaluation of a recursive definition. For instance, if *parent* were defined using the loose DML and *ancestor* were defined in the obvious way, then during the evaluation of the goal

?-ancestor (X, peter).

when the recursive clause was used, the goals to solve would be *parent(X,Z)* and *ancestor(Z,peter)*. Since X and Z are not instantiated, to solve *parent(X,Z)* the goal *retrieve([parent.father=X,parent.child=Y],true)* would have to be solved. We generalize this sort of situation to the case of any base relation being queried with the boolean condition set to *true*. We map the loose DML expression into the corresponding close DML expression but with all the attributes as variables. The built-in predicate *$whole_base_r* makes the appropriate tests to decide on the applicability of this transformation rule.

The second case is the one where intermediate queries have already been evaluated. Although this situation often arises during the evaluation of recursive queries, it is not exclusive to them. For these cases Educe keeps the results of intermediate queries until the whole query has been evaluated. These intermediate results are labeled with the generating query so that they can be matched to new queries.

The third case of importance occurs in conjunctive queries on a single relation. Again, this is a very common situation during the evaluation of recursive queries (top level). In this case the conjunctive query is first transformed into a normalized form, a logical simplification of the expression is performed, and then from this normalized form, an equivalent query in close form is generated and evaluated.

The last but certainly not the least major optimization step takes place during the transformation of loose form into close form and during the instantiation of variables at the top level. For both of these processes it is necessary to access the schema of the relation involved. These accesses are speeded up by maintaining the data base schema in buffers in main memory. In addition, whenever a tuple is retrieved from a base relation, a number of variables have to be instantiated. To do this, the list of values in the retrieved tuple has to be matched with a list of variables (typically, the variables in the projection part of a *retrieve*). The list of variables is normally shorter than the list of values and their sequences do not match. For instance, given the relation *employee* with attributes *name, plant, department* and the goal *retrieve([employee.plant=Plant, employee.name=Name],...)*, the first tuple retrieved might be *[john, munich, toys]*. Obviously, the order and the length of the lists *[Plant, Name] and [john, munich, toys]* do not match. In general, to match the two lists every time a new tuple is retrieved is unnecessary. The problem is avoided by, firstly, creating a bogus list of variables, say *[X1, X2, X3]*, then matching this list only once to the projection list in the *retrieve*. Finally, each time a tuple is retrieved from secondary memory, this bogus list is used to instantiate the real variables. A lot of unnecessary sorting is thus avoided.

The optimization steps described above form part of a new strategy of evaluation for queries expressed in the loose DML. We call it the *Educe method*. The improvements obtained by the application of this method go well beyond the recursive case. In fact, the *Educe method* always gives a performance close to the best of the other two methods (DBMs and Prolog) [3].

4. THE UNSOLVED PROBLEMS

Implementation dependent problems were discussed in the previous section. In this section we discuss what we see as more fundamental shortcomings of Educe, which we believe are inherent in any system constructed by the coupling/integration of Prolog and a DBMS.

A primary goal in the implementation of Educe was to provide users with a logical programming system for the construction of large KBMSs [18]. We wanted to keep the external appearance of this system as close to Prolog as possible. This was dictated by the early development of other prototype systems within our research group. These prototypes were all originally implemented in Prolog [5]. Thus, in the particular case of the close DML we wanted Educe to be transparent to Prolog users. Let us start by examining the extent to which this proved possible. The cases in which transparency was forfeited were:

1. Queries on facts stored in relations.
2. Compound terms.
3. Insertion/deletion rules.
4. Dynamic updating of Prolog programs.

5. Extended Prolog.

The Prolog user can use *asserta* to add a new fact or rule at the beginning of the clauses for a predicate, *assertz* to add them at the end, and 'assert' if it doesn't matter. For example if one writes in Prolog

> ?- asserta(r(1)).
> ?- assertz(r(2)).
> ?- assertz(r(3)).
> ?- assert(r(a)).
> ?- assert(r(b)).
> ?-r(X).

the answers 1, 2, 3 will be returned in order, but one does not know when the answers a and b will be returned, either before, after or amongst the other answers. In Educe, if a user wants to impose a specific ordering on any set of facts in the EDB they must be held as rules. This is because relations by definition are an unordered set.

Even if there is a unary relation *r* in the Educe database, the user cannot insert into it the fact *?- r(head(a))* because *head(a)* is a compound term. There is no database type *structure* in which to store Prolog structures. The reason for this is that the types provided by Educe are precisely the Ingres types. When an Educe relation is created, its definition is passed directly to Ingres. This avoids the more serious problem of having to maintain duplicate schemas, one in Prolog and one in the EDB. One possible solution that we have considered is to have intermediate Educe types which can map onto the underlying Ingres types. This can be implemented by adding a new system relation for this purpose to the EDB.

When a Prolog clause is asserted it can be either at the beginning of the clauses for a particular predicate (*asserta*), at the end (*assertz*), or its position may be unspecified (*assert*). To insert a rule into Educe it is necessary to give it a number (distinct from all the other rules for the same predicate). To retract a rule the user must either supply this number, or take the textual representation of the rule, append a full stop, convert it to an atom and then do the deletion from the database. Clearly insertion/deletion of facts and rules into the Educe database cannot be transparent since Educe must still support the standad semantics of assert and retract. However the only reason for requiring the user to assign a number to each rule is because it is used to force an order in the otherwise unordered relation *rulerel*. This is one more example of the contradiction between the procedural nature of Prolog and the unordered set concept of relational DBMSs.

The following point is not a question of transparency because it does not affect ordinary Prolog calls. However, we feel it is a related issue and hence it has been included here. The extension to Prolog in the close DML allows a call of

> ?- retr(r(A,B)),B>20.

for example, to be expressed as

?-retr(r(A,B>20)).

In general any comparison (>,>=,=:=,=\=,=<,<) can be expressed in the embedded form (characters are ordered on their ASCII representation). In order to gain the efficiency advantage that Educe provides for such queries, the user needs to use the extended syntax. Unfortunately, the syntax used by Educe in the extension is ambiguous. For example, to store two clauses for the predicate **r**, one in the relation *r*, and one in the rule relation (*rulerel*), we might add the fact *r(fred,20)* in the relation *r*, and the rule *r(X,50):- manager(X)*, in the rule relation. To make the close syntax transparent we can add an extra rule *r(X,Y) :- retr(r(X,Y))*. (In fact this could be done for every stored relation). Now the query

?- r(X,Y).

will succeed, returning {X=fred, Y=20} and {X=M, Y=50} for every manager M. Suppose the user wishes to take advantage of the extended syntax. He can ask

?- r(X, Y>10).

and this will call the subgoal

?- retr(r(X,Y>10)).

which will efficiently return the answer {X=fred,Y=20}. It will not, however, return {X=M,Y=50} for any manager M because the other rule for **r** has head *r(X,50)*, which will not unify with *r(X,Y>10)*.

Thus the user has to express the rule as:

r(X,Y) :- var(Y), !, manager (X), Y=50.
r(X,Y) :- integer(Y), !, Y=50, manager(X).
r(X,Comp) :- comparison(Comp,Y,Op,V),
    Y=50, call(Comp),
    manager(X).
comparison(X>Y,X,'>',Y).
comparison(X>=Y,X,'>=',Y).

to resolve this particular problem.

Two more problems worth mentioning are debugging and error handling. Usual Prolog facilities for debugging are just not satisfactory in the context of large knowledge bases. Just imagine trying to detect a loop involving a few thousand tuples in a relation which is used in a recursive definition of a predicate. Spying and tracing are too primitive for such a problem. The error handling problem is due to the different treatment and recovery techniques that the two components (Prolog and EDB) use. In Educe we tried to homogenize the situation by adopting the Prolog treatment. Unfortunately, this is not always possible, particularly in the case of the coupling, where there is little control over what is generated by the internal modules of the EDB.

5. CONCLUSIONS AND FURTHER WORK

In this paper, we have undertaken a critical revision of our work on Educe, a logic programming language for the development of KBMS. The version of the system here described has been successfully tested with the implementation of KB-2 [18]. We have also tested the performance of Educe with queries on reasonable large relations and several levels of recursion [3]. Response time for the first answer and for backtracking answers is never more than a few seconds. We consider that this response time is satisfactory. We do not expect any significant deterioration in performance on very large knowledge bases.

From the discussion on transparency, it is clear that although Educe is a close approximation to Prolog it is not an extended Prolog system which can store facts and goals on disc. However, it is indeed an efficient logic programming system for constructing large KBMS, as [17, 18] and [5] demonstrate.

Based on our past experience we have decided that the next version of Educe will move towards logic programming, and not towards Prolog. This is a departure from our adherence to Prolog. The new system Educe* is modular, uses new evaluation strategies and aims for completeness in the recursive case [17]. Also, in this new system we are relying much more upon our own access methods, so that we are less dependent on the underlying DBMS. All of this, plus the addition of structured data types, should resolve the main incompatibilities between the EDB and the Prolog component. Indeed, Educe* is no longer a marriage of these components, but a self standing system.

ACKNOWLEDGEMENTS

I would like to thank all the members of the Knowledge Base Group at ECRC, for their many helpful comments and discussions. In particular, I am grateful to Jean-Marie Nicolas, Mark Wallace and Michael Freeston for giving me so much of their time in discussions of this work.

REFERENCES

[1] Appelrath, H-J., Bense, H. and Rose, T. CPDB — A Data Base based Prolog System Incorporating Meta-Knowledge. Unpublished, September, 1985. ETH Zurich, Dept. of Computer Science, Switzerland.

[2] Bancilhon, F. and Ramakrishman, R. An Amateur's Introduction to

Recursive Query Processing Strategies. In Zaniolo, C. (editor), *Proc. ACM-SIGMOD 1986. International Conf. on Management of Data.* ACM, Washington, D.C., USA, May, 1986.

[3] Bocca, J. On the Evaluation Stategy of EDUCE. In Zaniolo, C. (editor), *Proc. 1986 ACM-SIGMOD International Conf. on Management of Data.* ACM, Washington, D.C., USA, May, 1986.

[4] Bocca, J. EDUCE — A Marriage of Convenience: Prolog and a Relational DBMS. In Keller, R. (editor), *Proc. '86 SLP Third IEEE Symposium on Logic Programming.* IEEE, Salt Lake City, Utah, USA, September, 1986.

[5] Bocca, J., Decker, H., Nicolas, J-M., Vielle, L. and Wallace, M. Some steps towards a DBMS based KBMS. In Kugler, H-J. (editor). *Proc. 10th World Computer Congress.* IFIP, Dublin, Ireland, September, 1986.

[6] Gallaire H., Minker, J., Nicolas, J. M. Logic and database: a deductive approach. *ACM Computing Surveys* 16(23), June, 1984.

[7] Gallaire, H. Logic Programming: Further Developments. In *Proc. 1985 Symposium on Logic Programming,* pages 88–96. Boston, USA. July, 1985.

[8] Henschen, L. J., Naqvi, S. A. On compiling queries in recursive first order databases. *Journal of the ACM* 31(1), January, 1984.

[9] Lozinskii, E. L. *Inference by generating and structuring of deductive databases.* Technical Report 84-11, Hebrew University, Department of Computer Science, June, 1984.

[10] Naish, L. *Mu-PROLOG 3.0 reference manual.* Melbourne University, Computer Sc., Melbourne, Australia, 1983.

[11] Sciore, E and Warren, D. S. Towards an Integrated Database-Prolog System. In *Proceedings First International Workshop on Expert Database Systems,* pages 801–815. Kaiwah Island, South Carolina, USA, October, 1984.

[12] Stonebraker, M. Inference in Data Base Systems Using Lazy Triggers. In *Proc. of the Islamorada Workshop on Large Scale Knowledge Base and Reasoning Systems,* pages 295–310. Islamorada, Florida, USA, February, 1985.

[13] Ullman, J. D. Implementation of logical query languages for databases. *ACM Transactions on Database Systems* 10(3):289–321, September, 1985.

[14] Vassiliou, Y., Clifford, J. and Jarke, M. Access to Specific Declarative Knowledge by Expert Systems: The Impact of Logic Programming. *Decision Support Systems* 1(1), 1984.

[15] Venken, R. The Interaction between Prolog and Relational Databases. *Unpublished,* Early, 1985. Report on ESPRIT Pilot Project 107.

[16] Vieille, L. *Recursive axioms in deductive databases: various solutions.,* Technical Report KB-6, E.C.R.C., Munich, Germany, May, 1985.

[17] Vieille, L. Recursive Axioms in Deductive Databases: The Query-

subquery approach. In *Proc. First International Conference on Expert Database Systems*. Charleston, South Carolina, USA, April, 1986.
[18] Wallace, M. G. *Reconciling Flexibility and Efficiency In a Knowledge Base Implementation*. Internal Report KB-8, European Computer-Industry Reserach Centre, Munich, September, 1985.
[19] Warren, D. H. D. Logic Programming and Knowledge Bases. In *Proc. of the Islamorada Workshop on Large Scale Knowledge Base and Reasoning Systems*, pages 69–72. Islamorada, Florida, USA, February, 1985.
[20] Zaniolo, C. Prolog: a Database Query Language for All Seasons. In *Proceedings First International Workshop on Expert Database Systems*, pages 63–73. Kaiwah Island, South Carolina, USA, October, 1983.

4

Extracting parallelism from sequential Prolog: Experiences with the Berkeley PLM

Wayne Citrin, Department of Electrical Engineering and Computer Science, University of California, Berkeley, California

1. INTRODUCTION

1.1 Aquarius

The Aquarius Project at Berkely [11] is an investigation of a new approach to computer organization in order to achieve high performance. The goal of this project is to discover the principles by which a machine can be organized such that it will concurrently execute both symbolic and numeric operations. The architecture is a heterogeneous, tightly coupled multiprocessor system with special processors to execute logic programming at the control level and numeric processors to execute array operations at the function level. Fig. 1 is an overall block diagram of the Aquarius heterogeneous MIMD machine.

The Aquarius group at the University of California, Berkely, has designed a special co-processor, the Programmed Logic Machine (PLM), as the first experimental processor that will lead to the Aquarius multiprocessor system. The PLM is a sequential machine executing an instruction set compiled from Prolog. Although we will see that the PLM provides better execution speed than other Prolog implementations, there are a number of possible avenues for further improvement. This paper will present our experiences with one such approach, the exploitation of the parallelism inherent in sequential Prolog programs. The conditions under which such parallelism appears, along with the methods for identifying these conditions, and the hardware support needed for parallel execution, will all be discussed here.

1.2 Prolog

Prolog is a logic programming language based on first order predicate calculus [18]. It is the implementation language being used in the Japanese Fifth Generation Computer Project [16] and the Berkeley PLM [12] and is used for a number of applications including expert systems and theorem

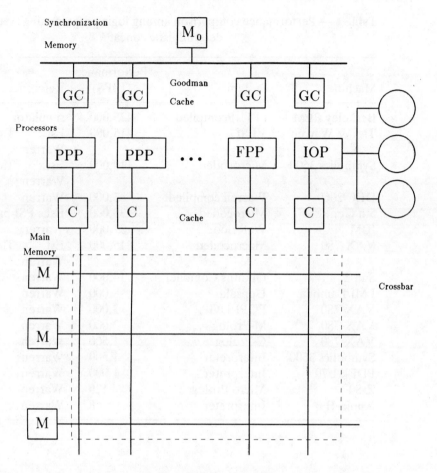

Fig. 1 — Aquarius architecture.

proving. It has also been employed as an implementation language for compilers [6, 20].

It is expected that the reader is familiar with Prolog. For more information on the language, see [8].

Among the drawbacks of Prolog up until now has been the relatively slow speed of Prolog implementations. Although not entirely satisfactory, the measure of execution speed in Prolog implementations is the logical inference per second (LIPS), where a logical inference corresponds to a procedure call. A more satisfactory measure is elapsed time. Under both measures, it can be seen that the sequential PLM is among the fastest Prolog implementations (Tables 1 and 2). To achieve additional speedup, we are exploiting several types of parallelism available in Prolog programs, namely AND-parallelism, OR-parallelism, and unification parallelism. Additionally, since the necessary information is available from analysis for the above parallelism, semi-intelligent backtracking is used for additional speedup.

Table 1 — Performance comparison among logic programming systems for deterministic concat

Machine	System	Performance (LIPS)	Reference
Berkeley PLM	(TTL)/compiled	425,000	simulator
Tick & Warren	VLSI	415,000	Est.: Tick & Warren
Symbolics 3600	Microcoded	110,000	Est.: Tick & Warren
DEC 2060	Warren compiled	43,000	Warren
5th Gen. PSI	Microcoded	30,000	Est.: PSI paper
IBM	Waterloo	27,000	Warren
VAX-780	Macrocoded	15,000	Est.: Tick & Warren
Sun-2	Quintus Compiler	14,000	Warren
LMI/Lambda	Uppsala	8,000	Warren
VAX-780	POPLOG	2,000	Warren
VAX-780	M-Prolog	2,000	Warren
VAX-780	C-Prolog	1,500	Warren
Symbolics 3600	Interpreter	1,500	Warren
PDP-11/70	Interpreter	1,000	Warren
Z-80	Micro Prolog	120	Warren
Apple-II	Interpreter	8	Warren

Table 2 — Benchmark timings (in ms)

Benchmark	Input	PLM	DEC-10 Comp.	C-Prolog	DEC-10 Int.
reverse	list30	4.4	54	250	1160
qsort	list50	4.9	75	410	1340
deriv	times10	0.38	3.0	25	76
	divide10	0.43	2.9	26	84
	log10	0.20	1.9	21	49
	ops8	0.25	2.2	22	64
serialize	palin5	3.2	40	240	600
query	—	17.3	185	4600	8900

2. THE BERKELEY PLM: A BRIEF DESCRIPTION

The Berkeley PLM [12] is an architecture based on the abstract instruction set presented by Warren [22] designed to efficiently execute compiled Prolog programs. In this section, we briefly present the memory organization and instruction set of the machine.

2.1 Memory layout

The PLM memory is divided into four areas. One, the code space, is read-only and data is not stored there by the PLM. The other three areas are used to store data.

The first data area is the heap, or global stack. It contains all structures created during the execution of the program.

The second data area is the local stack. It contains both environments and choice points. Environments contain those variables which may not be stored in machine registers because they must survive across calls. These variables are called "permanent variables" and the locations that they occupy in the environment are known as permanent registers. These variables are distinguished from temporary variables, which need not survive across calls and which may be kept in machine registers. Environments also contain information used to resume execution after a procedure is successfully completed, and to implement cuts and backtracking. A choice point, the other type of object which may be placed on the local stack, contains all the information necessary to restore the machine to a previous state upon backtracking.

The third data area is the trail, a stack which contains pointers to all variables which must be unbounded upon backtracking. Data is written automatically to the trail; assignments to it are not done explicitly.

2.2 Machine registers

The PLM contains a number of registers which define the machine state at any instant. The registers are:

P-   the program counter.
CP-   the continuation pointer. (Contains the return address to be used on successful completion of a procedure.
E-   pointer to the last environment pushed onto the stack.
N-   size of the current environment.
B-   pointer to the last choice point (for backtracking).
CF-   cut flag. Used to implement cut.
A-   top-of-stack pointer.
TR-   top-of-trail pointer.
H-   top-of-heap pointer.
HB-   heap backtrack pointer. (top of heap at last choice point.)
S-   structure point. Used when unifying structures.
X1,...,X8-
      argument and temporary registers.

An environment contains the following saved registers: CP, E, N, B, and the permanent registers. A choice contains the following saved registers: X1-8, E, CP, B, TR, H, N, and the address of the next clause should the current one fail.

2.3 Machine instructions

The PLM instruction set may be divided into two groups: unification and register-transfer instructions, and control instructions.

Unification/register-transfer instructions come in three clauses: **get, put,** and **unify.**

Get instructions are used to retrieve and unify a procedure's arguments. Some **get** instructions load a location with a value, and other will load a value if the target location is unbound, and otherwise unify the two values, causing a fail action **if** the unification does not succeed.

Put instructions simply load a location with a value. They are used to load argument registers in preparation for a call.

Unify instructions are like **get** and **put** instructions but act on elements of structures and lists.

Get, put, and **unify** instructions are further distinguished by the type of object which is to be unified with, or loaded into, a register. These distinguishing characteristics, known as annotations, are **variable, value, unsafe_ value , constant, structure, list, nil,** and **void**. The variable annotation is used at the first use of a variable in a clause and the value annotation is used for all subsequent uses. Unsafe_value is an annotation too specialized to discuss here, and the other annotations are self-explanatory.

The control instructions include instructions to set up choice points, perform branching depending on the type of value stored in a location, call and return from procedures, set up and remove environments, and implement the cut operator. The choice point instructions leave a choice point on the stack, so that if a failure occurs, the state may be restored and a new clause, pointed to by the choice point, executed. Branching instructions help narrow the choice of clauses to try, based on the type of first argument in the current call. A variety of call instructions are provided to handle normal calls, calls to built-in predicates, and final-call optimization. For more information on the PLM and the instruction set, see [12, 15].

3. STATIC DATA-DEPENDENCY ANALYSIS

Static data-dependency analysis (SDDA) is a pre-compile time technique developed by J-H Chang [3] to determine dependency relationships between subgoals in a given clause and coupling relationships between variables in a clause. Such information is then used to determine which subgoals in a clause may be executed in parallel (AND-parallelism), which subterms in a unification may be unified in parallel (unification parallelism), and which subgoals may be safely skipped upon backtracking (semi-intelligent backtracking). The ways in which this information may be used to achieve these optimizations will be described in subsequent sections.

SDDA begins when the programmer or user provides the analysis system with the source program and one or more **query entry modes,** each of which consists of the name of the predicate which will be called in the initial query, along with estimates of the coupling relationships between the subterms of the query predicate. Each of the subterms is identified as either a **ground** term (a constant or a structure all of whose arguments are ground terms), a **coupled** term (a non-ground term some part of which is bound to some other coupled term), or an **independent** term (a term which is neither ground nor coupled). In addition, coupled terms are partitioned into coupling classes, where all terms in any coupling class may potentially be coupled to each other. The analyzer is provided with one query entry mode for each query which may be expected to be made to the Prolog program, although if two queries have the same entry mode, only one of them need be provided. Table 3 gives some examples of query entry modes.

Table 3 — Query entry modes

Query	Entry mode
?-f(a,b)	entry(f,(g,g))
?-f(X,Y)	entry(f,(i,i))
?-f(X,Y,X,Y)	entry(f,(c_1,c_2,c_1,c_2))

This coupling information is then propagated throughout the program by a recursive algorithm explained below. The result is a set of entry and exit modes for each clause head and subgoal in the program, including an exit mode for the initial queries. Entry modes indicate the coupling status of the arguments of a clause head at the time of the call, or of the arguments of a subgoal just before it is attempted. Analogously, the exit mode of a clause head indicates the coupling status of its arguments at completion of the clause, while that of a subgoal indicates coupling status of the subgoal's arguments afer the subgoal is completed.

Since a clause may be called from many places and with many different possible variable couplings, it is possible to have many distinct entry and exit modes for each clause head and call subgoal. To simplify the analysis, only "worst-case" modes are computed. In comparing subterms, a ground term is "better" than an independent term, which is better than a coupled term. One mode is worse than another if the mode of at least one argument in the first is worse than the mode of its corresponding argument in the second, and none of the arguments in the first mode are better arguments in the second. The idea of a worse-case mode is useful because it will be seen that any parallelism scheduled using the worst-case mode will be correct even if the actual mode leads to better scheduling. It is possible that one mode may be neither better nor worse than another. In that case, a mode which is worse than both may be constructed by taking from each corresponding pair of mode arguments the worst one. Where two corresponding arguments are coupled terms from different coupling classes, the worst-case combines the coupling classes.

As an example of such a worst-case generalization, consider two modes entry $(f,(g,i,c_1,c_1,i))$ and entry $(f,(i,g,i,c_2,c_2))$. The worst-case generalization is entry $(f,(i,i,c_1,c_1,c_1))$.

As mentioned before, coupling information is propagated through the program using a recursive algorithm. For each query entry mode, each candidate clause called by the query is visited in turn. First the clause head entry mode is determined, then the entry mode of the first subgoal (if present) is determined. If the subgoal is a call, each candidate clause of the called procedure is then examined in the same way. When all clauses of the called procedure have been visited, a worst-case generalization is made of all the exit modes for that subgoal, the entry mode for the next subgoal is computed, and that called procedure is visited in the same way. When all the subgoals have been visited and their modes computed, the clause's exit mode is computed. When all clauses have been visited with the entry mode in question, a worst-case generalization of the exit modes is computed, which becomes the exit mode for that procedure, given the current worst-case entry mode.

To avoid redundant computation and infinite recursion, if a procedure is called with an entry mode which is equal to or better than the previous worst-case entry mode, the procedure is not re-examined. If it is called with an entry mode which is worse or unrelated, the procedure is re-examined with the new worst-case generalization.

As an example of SDDA, consider the list concatenation procedure:

concat([],L,L).
concat([X|L1],L2,[X|L3]):- concat(L1,L2,L3).

We will consider the case where concat is called with full instantiated lists in the first two arguments and an unbound variable in the third. This query entry mode is designated as query_entry(concat,(g,g,i)).

Visiting the first clause of concat, we get the entry mode entry(concat,(g,g,i)) and the exit mode exit(concat,(g,g,i)). Visiting the second clause with the same query entry mode, we also get the head entry mode entry(concat,(g,g,i)). (Because X is a ground term but L3 is independent, [X|L3] is an independent term.) The entry mode of the first subgoal is entry(concat,(g,g,i)) and since concat has already been visited with this entry mode, we know that the exit mode is still exit(concat,(g,g,g)). Thus, the query exit mode is query_exit(concat,(g,g,g)). In other words, when concat is called with ground terms in its first two arguments, the final argument will return a ground term.

The other product of SDDA is a data dependency graph for each clause.

Such a graph is directed and acyclic, and may have more than one root. The dependency graph is computed as follows:

(1) The **generators** of each variable in the clause are computed. The clause head is automatically a generator of each variable. In addition, any subgoal which contributes to the binding of an independent variable is considered a generator of the variable. Finally, any subgoal which contributes to the binding of a coupled variable is considered a generator of each variable in that coupling class.
(2) For each variable in a subgoal, the generator of that variable which most immediately precedes the subgoal is considered a **predecessor** of the subgoal. A subgoal may have several predecessors, one for each variable.
(3) For the clause head and each subgoal in the clause, a corresponding node is created in the dependency graph. If a' and b' are nodes in the graph corresponding to subgoals a and b, respectively, and a is a predecessor of b, put a directed edge from a' to b'. We say that b **depends on** a. Fig. 2 gives a dependency graph for the program solving the four-

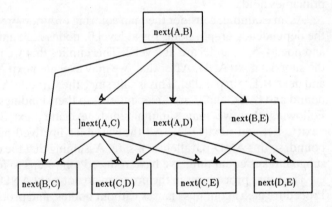

Fig. 2 — Dependency graph for color/5.

color map problem found in the appendix [4].

4. AND-PARALLELISM

4.1 Conditions

AND-parallelism occurs when two or more subgoals in the same clause are executed simultaneously. Fagin [14] has pointed out that the key problem in AND-parallelism is the possibility that two subgoals executed in parallel will generate conflicting bindings. When this is allowed, it is known as **unrestricted AND-parallelism**. The problem of resolving conflicting bindings in unrestricted AND-Parallelism is complex and not well-understood [10]. Fig.

3 shows a problem with conflicting bindings in unrestricted AND-parallelism if b and c are called in parallel. Somehow the conflicting bindings of X and Y must be resolved.

To avoid this problem, we have employed **restricted** AND-parallelism [4], in which no pair of subgoals may be executed in parallel if there is a possiblity that they may produce conflicting bindings. The way in which this is insured is through the use of dependency graph. Given a data dependency graph for a clause, two goals may be executed in parallel if neither is dependent on the other. If a goal b is dependent on a goal a, a must be executed before b. In terms of the dependency graph, a and b are independent if there is no path between the nodes corresponding to a and b. If we assign level numbers to the nodes in the dependency graph, so that the root nodes (nodes with indegree 0) are at level 0, and every other node is assigned a level number one higher than the level number of that node's highest-level ancestor, then a call subgoal may be executed only after all its ancestors are executed, and all subgoals in a given level may be executed simultaneously provided that all subgoals in the previous level have completed. This scheduling is subject to availability of processors, of course, but even with fewer processors than the largest number of nodes at any level, the same principles hold.

As an example, consider the map coloring program presented earlier. In the dependency graph, node 1 is at level 0, nodes 2, 3, and 6 are at level 1, and noes 4, 5, 7, and 8 are at level 2. This implies that we must first execute the subgoal next(A,B). After that, we may execute next(A,C), next(A,D), and next(B,E), in parallel. This is safe since the variable A has already been bound and C, D, and E are independent, so their bindings will not conflict. Following that we may simultaneously execute next(B,C), next(C,D), next(C,E), and next(D,E). All the variables involved have already been bound, so this AND-parallelism is safe. Assuming that the execution of each subgoal takes equal time, we have reduced nine steps to four steps.

The other procedures in the program do not yield AND-parallel savings. Procedure next1 contains clauses without bodies, and procedures next2 and next both contain single subgoal bodies.

Our techniques for AND-parallelism has advantages over Conery's [9] in that it requires no run-time checking, and needs no method for resolving conflicting references, as does his.

4.2 Hardware support

The actual details of the hardware extensions to the PLM to support AND-parallelism are currently being developed by Fagin [14], but a simplified description of these extensions may be given. The PLM **call** instruction, which is a traditional call, is replaced with a **call_p** (parallel call) instruction, which forks a process to perform the call, then proceeds to the next instruction. A sequence of these call_p instructions will create a group of AND-parallel processes.

In order to prevent subgoals from starting before their predecessors have finished executing, some sort of synchronization mechanism is necessary. In

```
?- a(Z,Z).

a(X,Y):- b(X), c(Y).
b(2).
c(3).
```

Fig. 3 — Conflicting bindings in AND-parallelism.

our extended PLM, each clause in which AND-parallelism occurs contains a **join table**, which is initialized to contain an entry for each level of the dependency graph with the number of nodes on that level. The call_p instructions take the appropriate index into the join table as an argument, and decrement that entry upon successful return. After forking a series of AND-parallel processes, the main process executes an instruction causing it to wait until the contents of the appropriate join table entry becomes 0. At that point, execution may proceed. In this way, synchronization is enforced.

Memory management is complicated by the introduction of AND-parallelism. AND-parallel processes should be able to access objects created before the processes were forked, but of course, they and any descendants must have independent environments. Thus, stacks reflecting the forking structure of the program must be available. In such "cactus" stacks, unlike conventional stacks where contiguous space is allocated for consecutive stack frames, frames must be allocated from a common pool.

5. SEMI-INTELLIGENT BACKTRACKING
5.1 Conditions
In addition to AND-parallelism, dependency graphs can be used to implement semi-intelligent backtracking. In conventional Prolog, if a call subgoal can invoke several candidate clauses, a choce point is created and the first candidate clause is executed. If this clause fails, a subsequent candidate clause is executed, the process being repeated until a clause returns successfully. It may happen that a clause returns successfully, and somewhere later in the program there is a failure. Backtracking may eventually return to the aforementioned call and the next candidate be tried. It it succeeds, execution will resume from this point. A major inefficiency in this type of backtracking is that trying a new clause at this choice point may not alter the values of any variables which would allow the clause that originally failed to now succeed. In such a case, all the work after the newly tried clause will be wasted. In Fig. 4, redundant work of this type takes place

```
?- a.
a:- b(X), c(3).
b(1).
b(2).
c(2).
```

Fig. 4 — Program with unnecessary backtracking.

when the second clause of b is tried. It will not cause c to succeed. It would be more efficient in this case to remove b's choice point and allow the clause to fail altogether. Semi-intelligent backtracking allows us to do this.

Chang [5] has identified three types of backtracking, for two of which there are practical compile time techniques for intelligent backtracking. These types of backtracking are:

(1) the case in which the backtrack is due to the failure of a goal during forward execution.
(2) the case in which the backtrack is due to the failure of a goal during backward execution (i.e., the goal in question was reached by backtracking from another goal, and itself failed), and where the initial failure occurred during execution of current clause.
(3) the case similar to #2 but where the initial failure occurred during some other clause.

As an example to type 1 backtracking, consider the map coloring program. Assume that the next(A,C) succeeds, and next(A,D) is then attempted and fails. Backtracking to next(A,C) can only change the binding of C, but neither A nor D. Therefore a new success of next(A,C) cannot possibly cause next(A,D) to succeed. Backtracking should go back to next(A,B), the closest subgoal which can possibly change the binding of an argument to next(A,D), in this case, A. In the case of type 1 backtracking, then, backtracking should go to the closest predecessor of the failed subgoal, as seen in Fig. 5.

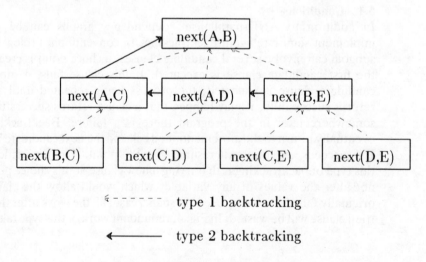

Fig. 5 — Type 1 and 2 backtracking for color.

Considering the same program, the following is an example of type 2 backtracking. Assume next(C,E) fails in forward execution. Type 1 backtracking will return to next(B,E). Suppose then that next(B,E) fails on re-execution. The backtracking from next(B,E) is type 2 backtracking. Type 1 backtracking would indicate that we return to next(A,B). However, we can safely backtrack only to next (A,D). This is because we cannot tell at this time whether the backtrack to next(B,E) was due to a failure in next(C,E) or next(D,E). We must therefore backtrack to the closest subgoal whose re-execution may solve the cause of failure, in this case, next(A,D). We determine the type 2 backtrack destination for a given subgoal by first calculating the set of all subsequent subgoals in the clause which may ultimately backtrack to the given subgoal (called the **backfrom** set). The type 2 backtrack destination is the closest preceeding subgoal for which there is a path in the dependency graph from that predecessor to some member of the backfrom set. In the above case, the backfrom set of next(B,E) is {next(C,E), next(D,E)}. There is a path from next(A,D), so next(A,D) is the type 2 backtrack destination. Figure 5 shows the type 2 backtrack destinations for all subgoals in the clause color. Note that subgoals which are leafs of the dependency graph have no type 2 backtrack destination.

In type 3 backtracking, the subgoal fails in backwards execution where the initial failure occurred outside the clause. Consider the example

$$r(W,Z):-u(W,Z), t(W,Z).$$
$$u(X,Y):-v(X), w(Y).$$
$$u(X,Y):-...$$

where X and Y are independent variables, v is the generator of X, and w is the generator of Y. Suppose that, at run time, t fails and backtracks to w, which also fails. Should backtracking go to v, which seems to have nothing to do with w, or to u, so that a new clause is tried? In this case, we must backtrack to v, which may change the value of W and cause t to succeed. In order to achieve intelligent backtracking in situations like this, it is necessary to combine dependency graphs of two or more clauses. This is complicated and has not been done in our scheme. Instead, we rely on naive backtracking in these situations.

It should be noted that in Porto and Pereira's intelligent scheme [19], intelligent backtracking may proceed across clause boundaries. However, the run time overhead necessary to implement the intelligent scheme offsets the advantages, as will be seen in section 8.

5.2 Hardware support

Details of the extended PLM environment for implementing semi-intelligent backtracking will not be presented here, but are available [3]. Basically, the code for each clause in which semi-intelligent backtracking is advantageous contains a backtrack table generated by the compiler. This has an entry for each subgoal in the clause, which indicates that subgoal's type 1 and type 2 (if it exists) backtracking destinations. A special instruction indicating the address of the backtrack table, is stored in the current (active) environment. Each call instruction contains an identifier for the subgoal being executed, namely the index into the backtrack table, which is stored in the new choice point. Upon backtracking, hardware flags indicate whether type 1, 2, or 3 backtracking has occurred, and the hardware takes the appropriate action.

6. OR-PARALLELISM

OR-parallelism is the parallelism achieved when multiple candidate clauses are executed for a given subgoal. Only one of the candidate clauses need succeed for the subgoal to succeed.

One of the difficulties in OR-parallelism is the handling of multiple bindings of the same variables. Several clauses may be activated by a single call, all of which may bind the same variables. Only when the calling subgoal chooses a candidate clause to "listen to," should any of the bindings become visible to the caller. Additionally, OR-parallel candidate clauses must not see each other's bindings. Although it has not yet been decided, it is expected that a variant of the hash window scheme [1] will be used.

A second difficulty with OR-parallelism is that it is of limited usefulness in deterministic programs. If a number of OR-parallel clauses are started, and the top-most one succeeds, then the work done by the others is wasted if that goal is never re-executed so that additional values are not returned. This can be a problem if the OR-parallel processes consume a large number of procesors, making them unavailable for necessary AND-parallelism. The solution is to assign a higher priority to AND-parallel processes than to the OR-parallel processes that they invoke, so that OR-parallelism only occurs when there is not enough AND-parallelism to fully occupy the processors.

7. UNIFICATION PARALLELISM

7.1 Conditions

Unification is the fundamental operation of Prolog. It has been shown that approximately 50% of the execution time of a typical Prolog program may be spent on unification [23], and that unification is therefore a prime candidate for optimization. Although Dwork [13] has shown that unification is a log-space complete problem for P, meaning that it is unlikely to have a fast parallel solution, a large subset of unifications, or parts of unifications, do have a parallel solution [7].

Clause and subgoal entry modes generated by static data-dependency analysis may be used to identify those unifications for which a parallel solution exists. When unifying a subgoal and a clause head, the unification operation can be divided into subunifications, each of which unifies a subterm of the clause head with its corresponding subterm in the subgoal. As in AND-parallelism, the major problem is preventing the generation of conflicting bindings. For example, when unifying the terms f(a,b) and f(X,X), we may attempt to unify a with X and b with X simultaneously. This will generate conflicting bindings, since the first unification will have bound X to a, and the second X to b. Resolving these conflicts, in this case by signaling unification failure, can be complex and offset any advantages gained by parallel unification. The solution is to schedule the unifications to be executed in parallel so that no two subunifications executed simultaneously can possibly generate conflicting bindings. The above unification must proceed sequentially. However, the unification of f(a,b) and f(X,Y), where X and Y are independent, may proceed in parallel. Likewise, in unifying f(a,b,c) and f(X,Y,X), where X and Y are independent, X may be unified with a at the same time that Y is unified with b. Only then may the unification of X with c be attempted. There may be several valid schedules for parallel unification of a pair of terms, although it is desirable to find a optimal one.

The rule behind parallel unification is that given the entry modes of the call subgoal and the clause head, two subterm pairs may be unified in parallel if the pairs do not share a coupled term. The problem is complicated by the fact that two terms which are not coupled to each other prior to unification may become coupled later due to another subunification. For example, in unifying f(X,X,Y) and f(Z,W,W), where W, X, Y, and Z are independent, the subunifications of X with Z and Y with W are initially independent and may be done in parallel. However, if the unification of X and W is done first, X is now coupled with W, and the first and third subunification pairs may not be done in parallel.

It is also possible for two subterm pairs which were initially coupled to be independent later on, and therefore unifiable in parallel if deferred. For example, when f(a,X,Y) and f(Z,Z,Z) are unified (X, Y, and Z being independent), all three pairs are coupled and may not be unified simultaneously. However, if the first pair, a and Z, is unified first, Z becomes the ground term a, and the second and third pairs are not coupled, so that they may then be unified in parallel.

Unification scheduling is a generalization of the resource-constrained scheduling problem [7, 17] and is NP-complete. However, a number of good heuristic algorithms exist to give near optimal results [7].

To improve the performance of parallel unification scheduling still further, several enhancements may be made to SDDA. Since SDDA does not concern itself with the elements of a structure, terms like X and s(a,b,Y) will both be considered independent terms (assuming X and Y are independent variables). Likewise, s(a,X) and s(b,X) are considered coupled terms, although SDDA does not report that they contain a ground term. Thus, in the unification of f(s(a,X),X) and f(s(a,Y),Z), SDDA will report that the two subterm pairs s(a,X)/s(a,Y) and X/Z are coupled and that they may not be unified in parallel. However, it is possible to unify the arguments of the first subterm pair in parallel. To get this additional information, it is necessary to modify SDDA to propagate information about the couplings of structures and their arguments throughout the program.

The second enhancement to SDDA concerns the improvement of worst-case estimates. Given two possible entry modes for a call, one of which is worse than the other, the worse mode will yield a worse (i.e., longer) unification schedule. If only the worst-case mode is used, all calls with a better mode will receive a non-optimal (although safe) unification schedule. In order to achieve better worst-case estimates, a source transformation technique called procedure-splitting is used. If a clause is called from two places in the program with two distinct entry modes, the called clause is split into two distinct clauses each of which is called from one of the calling sites. Since the clauses are now separate and distinct, the entry modes of the two subgoals need not be combined into a worst-case generalization. Both clause head unifications may be given optimal unification schedules. Procedure-splitting may be incorporated into the static data-dependency analyzer.

As an example of procedure-splitting, consider the following program segment:

?-a(X,X). % query
a(X,Y):- a(Y,Z). % clause

The worst-case entry mode for a is entry(a,(c_1,c_1)) (from the query), although a is also called with two independent arguments (by way of recursive call). The worst-case estimate yielded by SDDA would have the arguments coupled. Scheduling according to this entry mode would require that the unifications of the two arguments of a be done sequentially, although when a is called recursively they may be done in parallel due to the independence of the arguments. To narrow the worst-case estimates and expose the potential parallelism, we split clause a into two clauses, a1 and a2, to be called when its arguments are coupled or independent, respectively. The transformed program would then look like this:

?-a1(X,X).
a1(X,Y):- a2(Y,Z).
a2(X,Y):- a2(Y,Z).

(Note that the calls in the bodies of the clauses are always to a2 because the arguments are independent.)

a1 still is executed with two coupled arguments; they must be unified sequentially. However a2 is always called with two independent arguments; its arguments may always be unified in parallel.

7.2 Hardware support

In the sequential PLM, unification instructions are associated with the head of the clause being called. That is, the clause is called, the called clause is entered, and the arguments of the clause head are unified with those of the called, one symbol at a time. For example, in the clause head

f(a,X,s(b,Y,X)):-

code that looks like this is generated:

label f:

```
get_constant a,X1    %Xn=nth argument
get_variable X,X2
get_structure s/3,X3
unify_constant b
unify_variable Y
unify_value X
...
```

Each unification instruction corresponds to a symbol: a constant, variable, or structure functor. Each instruction, then, corresponds to a subterm pair to be unified. When a parallel schedule of subterm pairs is created, a corresponding block of parallel head unification code can be created for any compiled PLM program. Multiple parallel unifications are indicated by wide horizontal unification instructions. While the format has not yet been determined, one possibility is a continuation flag in each instruction which, when set, indicates that the following instruction is to be executed in parallel with the current instruction, or inversely, when set indicates that the following instruction is to be done sequentially (not in parallel). Also, for structure and list element unifications, an argument is needed in the unify instruction to indicate the index of the argument. In the sequential version, instruction order indicates the argument index.

8. RESULTS

Although our results have not yet been simulated on a large number of benchmarks, preliminary results are encouraging. Fagin [14] has simulated execution of the quicksort program from the Warren Benchmark Set [21] which has been modified to produce more AND-parallelism (Fig. 6). The

```
quicksort([X|L],R,R0):-        %arg1=input, arg2=output,
    partition(L,X,L1,L2),      %arg3=[] in initial query
    qsort(L2,S2,R0).
    qsort(L1,R,[X|S1]),
    S2=S1.
qsort([],R,R).
partition([X|L],Y,[X|L1],L2):-
    X<Y, !,
    partition(L,Y,L1,L2).
partition([X|L],Y,L1,[X|L2]):-
    partition(L,Y,L1,L2).
partition([],_,[],[]).
```

Fig. 6 — quicksort.

simulator models a parallel PLM with the assumption that all parallel instructions take an equal amount of time and that processors execute in lockstep (an admittedly unreasonable assumption).

The results are shown in Fig. 7 (from [14]) for input lists of 64, 128, and

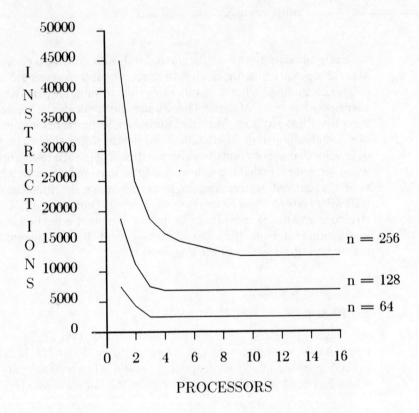

Fig. 7 — Performance improvement through adding processors.

256 elements. Note that the number of parallel instructions executed decreases dramatically as the number of available processors is increased, but tails off rapidly. As might be expected, the change from 1 to 2 processors is the greatest. This corresponds to the introduction of AND-parallelism.

The theoretical execution time for sequential parallelism is $Q(n)=n \log n$, while for the parallel version, it is $Q(n)=2n-1$ (assuming balanced partition trees). This leads to an expected speedup of $O((\log n)/2)$. We find that the actual speedup is fairly close to the predicted speedup (Table 4).

The difference between expected and actual speedup is chiefly due to an unbalanced partition tree.

Chang [3] has tested semi-intelligent backtracking on a modified PLM simulator and compared the results to those of Bruynooghe and Pereira's intelligent backtracking [2]. Simulations were done on the abovementioned

Table 4 — expected vs. actual speedup

n	Expected	Actual
64	3	2.3
128	3.5	2.5
256	4	3.33

map coloring program, and on a simple database query program (Table 5). In the case of the PLM-simulated programs, total number of instructions and number of failures are counted. For intelligent backtracking, an unspecified "percentage change in performance" is included. For the map coloring program, easy and difficult problems were tested. In all cases, improvements were found by using semi-intelligent over naive backtracking. Where the run-time scheme does have an advantage, it is due to the presence of type 3 backtracking in the program.

Note that '−' denotes performance improvement and '+' denotes degradation. Where performance degradation exists, it is due to run time overhead offsetting the decrease in the number of failures.

In order to test unification parallelism, SDDA (the non-enhanced version) was performed on a set of benchmark programs, and the data used to schedule the unifications and generate horizontal parallel unification instructions. The ratio of sequential to parallel unification instructions (Table 6) ranged from 2.1:1 to 3.6:1, and would probably be higher had enhanced SDDA and procedure splitting been used. Actual speedup measurement in terms of machine cycles has not yet been done, but should be done soon. It is hoped that the speedups for the unification instructions will be comparable.

9. CONCLUSION

The results obtained so far for extracting parallelism from sequential Prolog programs are encouraging. As work proceeds on design of the actual parallel PLM hardware and that design is simulated more precisely, better information on the speedup will be gathered. In addition, simulation and analysis of larger Prolog programs, such as the Berkeley PLM Prolog compiler, will provide more valuable and practical information. It is expected that a simulator will be ready by December 1986, with construction of the machine beginning the following year.

ACKNOWLEDGEMENT

The Aquarius Project, of which this work is a part, is largely a group effort. The author wishes to express his gratitude to the other members of the group: Alvin Despain, Tep Dobry, Barry Fagin, Philip Bitar, Jong-Herng

Table 5 — Naive vs. (semi-) intelligent backtracking

	Naive		Semi		Semi vs Naive		B&P vs Naive
Benchmark	Instr	Fail	Instr	Fail	Instr	Fail	
query	1493	289	1258	186	−16%	−35%	−20%
map-hard	2484893	714808	2272	371	−99.9%	−99.9%	−99.7%
map-easy	894	200	913	185	+0.7%	−7.5%	+63%

Table 6 — Unification parallelism

Benchmarks	Parallelism
query	2.1:1
mu	2.1:1
serialize	2.7:1
deriv	3.6:1
nreverse	2.3:1
quicksort	2.2:1

Chang, Wen-Mei Hwu, Steve Melvin, Michael Shebanow, Vason Srini, Yale Patt, and Robert Yung. We would also like to thank Doug DeGroot of IBM research for his suggestions and encouragement.

This work was partially sponsored by Defense Advance Research Projects Agency (DOD) and monitored by Naval Electronic System Command under Contract No N00039–84–C–0089 and Contract No. N0039–83–C–0107. Support from the Advanced Computer Architecture Laboratory of the Lawrence Berkeley Laboratory is gratefully acknowledged. Part of the work was sponsored by the California MICRO program. Thanks are also due to the NCR Corporation for their generous contribution of equipment, and for funding part of the work done under the project.

APPENDIX: A MAP COLORING PROGRAM

```
% program to color a map with five regions: A, B, C, D, E.
% Following adjacencies: (A,B), (A,C), (A,D), (B,C), (C,D),
% (B,E), (C,E), (D,E).

color(A,B,C,D,E):-
    next(A,B), next(A,C), next(A,D), next(B,C),
    next(C,D), next(B,E), next(C,E), next(D,E).

next(X,Y):- next1(X,Y).
```

next(X,Y):- next2(X,Y).

next1(green,red).
next1(green,yellow).
next1(green,blue).
next1(red,blue).
next1(red,yellow).
next1(blue,yellow).

next2(X,Y):- next1(Y,X).

REFERENCES

[1] P. Borgwardt, Parallel Prolog Using Stack Segments on Shared-Memory Multiprocessors, *Proceedings of the Symposium on Logic Programming — 1984,* February 1984.

[2] M. Bruynooghe, and L. M. Pereira, Revision of top-down logical reasoning through intelligent backtracking, Report 3/80, Universidade Nova de Lisboa, and Katholieke Universiteit Leuven, 1981.

[3] J.-H. Chang, High Performance Execution of Prolog Programs Based on a Static Data Dependency Analysis, Ph.D. Thesis, University of California, Berkeley, October 1985. Available as Tech. Report UCB/CSD 86/263.

[4] J.-H. Chang, A. M. Despain, and D. DeGroot, AND-Parallelism of Logic Programs Based on a Static Data Dependency Analysis, *Proceedings of Spring 1985 Compcon,* San Francisco, March 1985.

[5] J.-H. Chang, and A. M. Despain, Semi-Intelligent Backtracking of Prolog Based on a Static Data Dependency Analysis, *Proceedings of the Symposium on Logic Programming, 1985,* Boston, July 1985.

[6] Wayne Citrin, Peter Van Roy, and Alvin M. Despain, Compiling Prolog for the Berkeley PLM, *Proceedings of the Hawaii International Conference on Systems Sciences — 1986,* Honolulu, HI, Janurary 1986.

[7] W. Citrin, Parallel Unification Scheduling in Prolog, Ph.D. Thesis, University of California, Berkeley. Expected December 1986.

[8] W. F. Clocksin and C. S. Mellish, *Programming in Prolog,* Springer-Verlag, New York, 1981.

[9] John S. Conery, The AND/OR Model for Parallel Interpretation of Logic Programs, Ph.D. Thesis, Dept. Information and Computer Science, Univ. Calif., Irvine, 1983.

[10] D. DeGroot, Restricted AND-parallelism, *Proceedings of the Int'l Conf. on 5th Gen. Comp. Sys.,* Tokyo, November 1984.

[11] T. P. Dobry, Jung-Herng Chang, Alvin M. Despain, and Yale N. Patt, Extending a Prolog Machine for Parallel Execution, *Proceedings of the Hawaii International Conference on Systems Sciences — 1986,* Honolulu, HI, January 1986.

[12] T. P. Dobry, Y. N. Patt, and A. M. Despain, Design Decisions

Influencing the Microarchitecture for a Prolog Machine, *Proceedings of the 17th Annual Micropgramming Workshop*, New Orleans, November 1984.
[13] C. Dwork, P. C. Kanellakis, and J. C. Mitchell, On the Sequential Nature of Unification, *The Journal of Logic Programming*, vol. 1, no. 1, pp. 35–50, June 1984.
[14] B. Fagin, *A Parallel Execution Model for Prolog: A Research Proposal*, University of California, Berkeley, April 1986. Internal memo.
[15] B. Fagin, The Berkeley PLM Instruction Set, Technical report, University of California, Berkeley.
[16] K. Furukawa and T. Yokoi, "Basic Software System," *Proceedings of International Conference on Fifth Generation Computers*, Tokyo, Japan, November 6–9, 1984.
[17] M. R. Garey and D. S. Johnson, *Computers and Intractibility: A Guide to the Theory of NP-Completeness*, W. H. Freeman, San Francisco, 1979.
[18] R. A. Kowalski, *Logic for Problem Solving*, North-Holland/Elsevier, New York, 1979.
[19] L. Pereira and A. Porto, Selective Backtracking, in *Logic Programming*, ed. S. A. Tarnlund, pp. 107–114, Academic Press, London, 1982.
[20] P. Van Roy, A Prolog Compiler for the PLM, Masters' Report, University of California, Berkeley, August 1984.
[21] D. H. D. Warren, Logic Programming and Compiler Writing, *Software Practice and Experience*, vol. 10, no. 2, pp. 97–126, February 1980.
[22] D. H. D. Warren, An Abstract Prolog Instruction Set, Technical Note 309, Aritificial Intelligence Center, SRI, Menlo Park, CA, October 1983.
[23] N. S. Woo, A Hardware Unification Unit: Design and analysis, *Proceedings of the 12th Intl. Symposium on Computer Architecture*, New Orleans, June 1985.

5

DEDALE: an Expert System in VM/Prolog

Philippe Dague, Philippe Devès†, Zeina Zein and Jean Pierre Adam IBM France, Scientific Centre, Paris, France

† Electronique Serge Dassault, Paris, France.

INTRODUCTION

DEDALE is a troubleshooting Expert System for analog electronic circuits. It is developed jointly by the Paris Scientific Centre and the Electronique Serge Dassault company (ESD).

Today, localization of defaults in digital circuits is largely automated through the use of efficient algorithms. Unfortunately, this is not the case for analog circuits, although, basically, their functioning (and malfunctioning) can be accounted for by known physical laws.

Although one can expect to use as much modelization as possible, based on those physical laws, recourse to the Expert System formalism is felt necessary, both to take benefit of know-how as formulated by experts, and to exercise control on the reasoning process of the system.

DEDALE is an attempt to mix heuristic knowledge of expert diagnosticians, as found in the first generation, shallow reasoning Expert Systems (like EMYCIN), to knowledge of the underlying structure and function of the circuit.

The expected advantage is to produce a deeper model of reasoning, taking advantage of physical theory at least at the local level of a particular component and function: this should result in a better guidance of the diagnostic process and should enable the system to produce an explanation based on causality.

KNOWLEDGE REPRESENTATION

With regard to typical medical diagnosis, analog electronic diagnosis present significant differences. The objects themselves are highly structured and an appropriate representation system for factual knowledge (here a Frame system) must be used. There are many different types of circuits, and the system does not possess an a priori knowledge of a particular one. Like a human expert using his general knowledge on the schema he is discovering, the system must conduct its reasoning on a representation of the particular circuit, using Expert System rules general enough to be usable in all kinds of situations. The system must also be able to use partial modelling of functions and components when deemed necessary.

The resulting Expert System includes the following major components and functions:

— Representation of all the analog electronic concepts useful for the debugging, as generic objects (components, higher level functions, etc).
— Representation of the actual defective circuit as instantiation of those generic objects.
— Representation of the troubleshooting expertise as a set of rules for heuristic, choice, reasoning strategy and procedures
— The inference system and the Frame system.

The overall architecture is derived from BSM [1], a previous system for medical application. The dual representation scheme of BSM, Frames for factual knowledge and Rules for inferential and control knowledge has proven even more appropriate for the analog circuit domain.

Knowledge about the circuit (factual knowledge)
The basic functional and structural elements (see below) are represented as generic objects in the Frame system. A description language with a precise syntax defines the specific circuit to be debugged (see Appendix I). This description is processed to build, through *Frame instantiation*, the internal description of the circuit (Fig. 1).

```
frame(B1, isa, val, [hybrid-block,amplifier]) .
frame(B1, included-in, val, C) .
frame(B1, composed-of, val, [R2,R5]) .
frame(B1, test-points, val, [t1,t2]) .
frame(B1, position(C), val, r(13,53)) .
frame(B1, size, val, rectangle(2,2,72,117)) .
```

Fig. 1 — Extract of the internal description of a circuit resulting from Frame instantiation.

Fig. 2 gives a graphical representation of a real circuit as it appears on the screen of the 3279.

Structural representation
The structural representation attempts to reproduce all physical and visual information useful for the debugging process: *spatial concepts*: physical layout, size, proximity, position of test points; *technology concepts*: connections and electronic components; *topological concepts*: connectivity.

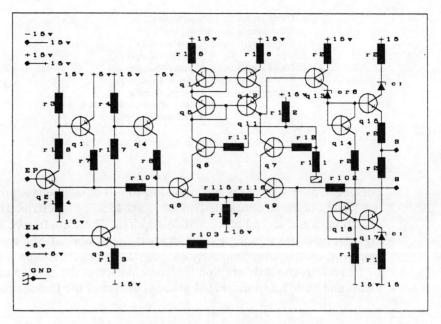

Fig. 2 — Representation of a real circuit on the 3279 screen.

The basic elements of this knowledge are:

- (structural-)**block**: physical entity having a position, a size, various technological attributes, test-points, and an internal structural decomposition in terms of block, atom, node.
- **Atom**: block without further decomposition useful for the debugging. An atom is the minimal entity which can be suspected. Generally, to an atom is associated a function.
- (internal-)**node**: physical entity associated to a set of physical connections which link points at the same potential, inside a block. It has technological attributes, test-points, and a structural decomposition in terms of (internal−)node, link.
- **link**: corresponds to an electrical link between test-points. As any atom, a link is a minimal entity which can be suspected of malfunctioning during the diagnostic process.
- **test-point**: physical point where electrical measurements can be made. In general, they are associated to a node.

Fig. 3 gives an extract of the Prolog predicates representing the frame prototype for the structural aspect of a resistor.

Functional representation
The functional representation of the circuit is done with the viewpoint of the diagnostician, rather than that of the designer. It is expressed in terms of a hierarchy of **functions** where the leafs are basic electronic functions corres-

```
frame (resistor, isa, val, atom).
frame (resistor, priority, val, 1 ).
frame (resistor, tolerance, domain ,
            [ between, 1.E-5, 0.2 ]).
frame (resistor, max-power, domain ,
            [ between, 1.E-2, 5 ]).
frame (resistor, technology, domain ,
            [ metal-layer, thin-layer, carbon-layer,
             thick-layer, coiled ]).
```

Fig. 3 — Extract of the prototype for a structural resistor.

ponding to atoms, or **nodes** corresponding to sets of interconnected (internal-)nodes, of the structural description. Each function is described by a set of attributes and several functional descriptions. There are frames describing basic functions such as: resistor, capacity, inductance, diode, transistor, ampli-op, comparator, amplifier, etc . . . . The bridge between the functional and structural description is a consequence of the overlap between atoms and basic functions. Fig. 4 gives an extract of the Prolog predicates

```
frame (resistor( [*N+,*N-,*R] ), isa, val, function).
frame (resistor( [*N+,*N-,*R] ), nodes, val, [*N+,*N-]).
frame (resistor( [*N+,*N-,*R] ), value, val, *R).
frame (resistor( [*N+,*N-,*R] ), type, val,
                  [resistor,impedance,passive-block]).
```

Fig. 4 — Extract of the prototype for a functional resistor.

representing the Frame prototype for the functional aspect of a resistor.

Knowledge about the troubleshooting process (inferential knowledge)
There are basically two types of knowledge: Experience (heuristic) knowledge and "exact" knowledge: physical laws in general, and modelling of physical devices. Some processes like structural analysis can also be algorithmically described. Various knowledge representation mechanisms are used in DEDALE: The "production rules" encode the heuristic knowledge, and part of the physical knowledge in the sense that appropriate predicates are activated during rule premises evaluation. Procedural knowledge is found in procedural attachment of frames (computation of Kirchhoff laws for example) and in Prolog programs activated outside the inference loop (such as structural analysis) or in special Prolog predicates activated during premises evaluation of rules.

Experience (heuristic) rules
Starting from initial information, experience rules generally formulate hypothesis about faults and possible causes. They result from the "heuristic" knowledge of experts.

Example: If there is a power supply failure at a node, then, for any block connected to that node and containing a non-validated zener diode, suspect a direct-polarization of that diode (Fig. 5).

```
IF    breakdown(power_supply, *N)
AND   *N ; linked-block ; *B
AND   EXEC( presence(*B, zener, *Z))
AND   NOT validation(*Z)
THEN  suspect(zener, *Z, direct_polarization)
```

Fig. 5 — Example of a heuristic rule.

Rules about functioning/malfunctioning

These rules attempt to prove that a function (suspected in a previous step) is working correctly or conversely, is malfunctioning. They are generally based on knowledge about physical laws or functions of devices (like transistors) but are not a thorough confirmation of the functioning/malfunctioning.

Example: If the absolute value of the voltage between the base and the emitter of a transistor is between 0.5 and 0.7 and if the intensity through the collector is close to zero, then the transistor is in open circuit, and the cause is a failure of the junction collector-base (Fig. 6).

```
IF    *base ; ddp(*emitter) ; *VBE
AND   EXEC( *U := abs(*VBE) & between(0.5, *U, 0.7))
AND   *collector ; intensity ; close-to ; 0
THEN
      malfunction(transistor(*base,*emitter,*collector),
                            *T, open-circuit)
AND   break-down(open-junction-collector-base, *T)
```

Fig. 6 — Example of a rule about a malfunctioning.

Rules about choice of hypothesis

They are a component of the diagnostic strategy: in presence of a set of hypothesis, they choose to validate one of them.

Example: If a function is suspect, select a sub-function to test, in accordance to some criteria such as confidence level attached to structural blocks. If a function is invalid, and if an hypothesis on a sub-function is formulated, attempt to validate this hypothesis. A node is a priori supposed to be valid, except reconsideration in particular situations.

Procedural knowledge

Influence analysis

A function influences another function if there is an electrical path between them. Influence analysis determines which higher level function might account for the faulty output, based on electrical path analysis. It is used to reduce the search of hypothesis to be examined at a given hierarchical level.

Priority analysis

A confidence level is associated to structural elements and can be modified as a result of the experience gained by the system. Influence information and confidence level are used by rules about choices.

Calculation of electrical parameters

It is sometimes possible to compute or deduce values rather than measuring them. DEDALE attempts systematically to apply Kirchhoff laws at node to compute voltage and current. When this is not possible, DEDALE currently asks the value of the current. Unfortunately, current measurement are generally not possible in real situations, because it is an operation which can be destructive or which modify the behaviour of the circuit by derivation of most of the current to the measuring device.

Future developments will attempt to compute currents from voltage measurement, and hypothesis about the correct functioning of the components involved in the computation. Inference might also be necessary to conduct this determination. These hypothesis might eventually be invalid, and for such a calculation to be valid, the system must be capable of dependency backtracking to formulate another hypothesis and do the computation again, should contradiction arise.

DIAGNOSTIC PROCESS

The following steps are executed during the diagnostic process:

- Description of the specific circuit to debug. The description language (Appendix I) is compiled into a set of terminal frames via instantiation of prototypes.
- Description of the observed default: list of faulty outputs and associated problems (absence of signal, wrong frequency, low voltage) and general problems associated to the circuit (temperature, high current consumption). These data are used by the heuristic rules. They are provided by the user in answer to questions resulting from triggering procedural attach-

ments, as a result of the evaluation of the premises of rules in the Frame system. This questioning mechanism is activated during the whole process of rule evaluation, and entirely handled by the frame prompting subsystem.

Ultimately, the voltage measurements will be automated, and the prompting subsystem will directly control the testing devices, thanks to the knowledge of the positions of the test-points.
- The influence analysis results in a set of potentially suspect high level functions.
- The diagnostic proceeds according to the following cycle:

  — Forward chaining on experience rules. This result in hypothesis about malfunctioning. They are expressed as Prolog terms such as:

    fault(capacity([N5,N8]), C3, short-circuit)

    and added as Prolog axioms in the Prolog data base. These hypothesis identify the function, the object, the possible cause.
  — Backward chaining on choice rules to select a single hypothesis based on priority and confidence coefficients. In the absence of hypothesis, the system will attempt to validate not yet validated functions.
  — Backward chaining on rules about functioning/malfunctioning to try to validate hypothesis. According to the presence or absence of a precise possible cause, the process tries to validate the malfunctioning first, or the functioning first. A record is made of the validated/non-validated elements in the fact data base.
- During these backward chaining phases, information on the circuit (resulting from measurement or question asked to the user) is acquired and recorded in the Frame system, and results (validation of failure to validate functions) are recorded in the fact data base, and will be used in the following cycles. This cycle stops if an atom, a link, are recognized as "bad" (i.e. the default is localized) by a rule, or if all functions are validated, in which case the debugging process fails to localize the problem and can only propose to replace the non-validated atoms. Otherwise, the inference loop is started again with the forward chaining on choice rules.

IMPLEMENTATION IN VM/PROLOG

BSM was implemented in YKT/LISP. DEDALE is entirely implemented in VM/Prolog. This implementation language has been chosen deliberately for all our developments in Artificial Intelligence at Paris Scientific Center. Our objective is to investigate, on real and complex problems of Knowledge processing, the applicability of the Logic Programming paradigm under VM/Prolog.

Prolog appears as a very valuable implementation language for Expert Systems. Its sound foundation on predicate logic, its powerful operation of unification on trees and its resolution mechanism makes it a very high level programming language, allowing to concentrate on main issues of Expert System programming rather than on low level technicalities. It provides significant productivity improvement over other languages, including Lisp, while keeping performance at least equivalent to Lisp implementations. Frame systems, inference engines and object oriented paradigm like SPOOL [2] are readily implemented in Prolog while a significant part of the search and matching mechanism usually found in those systems are handled by the interpreter, whereas they must be entirely programmed in other languages.

The underlying Expert System comprises an inference engine (backward and forward) and a Frame system very similar to the one found in BSM [1], including procedural attachments. As an exercise, we used a modified version of SPOOL to represent graphically the image of the circuit and the GDDM interface to display it. The procedural knowledge is implemented as Prolog predicates acting on the representation.

The inference system
A detailed description of the DEDALE system will be found in [3].

Syntax of rules
The syntax provides for complex expressions with operators such as $<;>$, AND, OR, AMONG, IMPLY, NOT (see Appendix II). They are infix predicates with appropriate priorities. The variables are Prolog variables. Rules can be grouped in packages and can be loaded and unloaded dynamically. Rules are Prolog terms. During loading, they are "compiled" into Prolog axioms, with two different representation (forward, backward). This approach allows use of all the powerful Prolog facilites: unification, backtracking, dynamic creation and manipulation of Prolog programs by Prolog.

Forward chaining
Forward chaining is generally made by saturation of the fact data base. The premises of rules are evaluated either into the fact data base (unification inside the Prolog data base), into the Frame representation (there is a set of standard predicates) or upon explicit execution of arbitrary Prolog predicate. The action part of rules modifies the fact data base, instantiates or adds values into the Frame system. or executes arbitrary Prolog predicate. Generally, the triggering of the procedural attachments of the prompting subsystem is inhibited during this phase.

An optimization ensures that only rules referring in their premises to previously modified part of the fact data base are triggered. This optimization can be extended to the Frame system. The control is made by an examination of the conclusions of the form "<fact>" or "NOT <fact>". This simple technique works only if the actions:

EXEC(<prolog_goal>)

do not introduce side effects in the fact data base or in the frames. If this is the case, the control should be made at the level of fact data base and would be expensive.

When the forward chaining system works in the open world mode, a coherence check (check for no contradiction) is made. The coherence inside the Frame system is maintained internally via specialized procedural attachment. Other control facilities are implemented to stop the chaining upon certain circumstances.

If free variables exist in the rule premises, several activations are generally possible, giving rise to generally different actions. They are differed and executed at the end of rule activation to avoid problems with side effects.

Backward chaining

A backward rule is triggered when a goal is started, in general by the evaluation of the premise of another rule. The possible triggers are the possible conclusions of rules:

<fact> NOT <fact> or
<object> ; <attribute> ; <value>

but not the   EXEC(<prolog_goal>).

The default strategy is to use the inference engine of Prolog to handle this backward chaining (rules are complied into Prolog clauses). If there are free variables in the goal, several triggerings are possible, and are handled by the Prolog backtracking.

Other strategies are possible and will be implemented. For instance, re-ordering the premises avoid to ask question or undertake expensive measurements if it can be established beforehand that the rule will fail. Re-ordering the rules according to some criteria (such as the cost of premise evaluation) insures that easier tests are made first.

To implement such strategies, knowledge about rules will be used by meta-rules. In Prolog, it is particularly easy to implement such meta-rules: rules are in fact implemented as Prolog terms, they can be manipulated as data objects through unification and then be used as Prolog predicates. Notice that Lisp allow also to represent programs as data objects, and allows the writing of programs to dynamically create and modify other programs. Lisp and Prolog can "talk" about themselves.

Notice also that whilst this powerful property makes implementation of meta rule facility easy, the underneath theoretical foundation of Expert Systems offering such facilities are not so clear, and should be sought for in non-classical logics underlying the inference system. Many theoretical results remain to be found. See [4] for a discussion.

FUTURE DEVELOPMENTS

The system will be really usable when electrical current calculation and hypothetico–deductive reasoning with dependency backtracking will be implemented.

Two other improvements will contribute to make the system really usable in an industrial environment: automated measurements, and handling of multiple faults in circuits.

We think that for inference systems in highly structured domains such as DEDALE, the use of reasoning strategies is absolutely essential. Human experts explain their approach in terms of strategies, plans, and cannot provide hundreds (or thousands) of debugging rules to be processed blindly by an inference system.

We hope to introduce high level reasoning strategies involving formalization of decision making and building of diagnostic plans.

We think also that the debugging process could be oriented by the human expert.

Another hope is that this approach incorporates methods of qualitative reasoning about physical systems [5]. One possible application would be to assess, through qualitative simulation of physical devices, to what extent the established default explains the observed misbehaviour.

REFERENCES

[1] Adam, J. P., Fargues, J., Pagès, J. C.: *BSM Project. The Expert System.* Internal Research Report F054, IBM Paris Scientific Center. 1984.

[2] Fukunaga, K., Hirose, S., Van Dam, E.: *SPOOL, User's manual.* IBM Japan Science Institute (in French).

[3] Dague, P., Devès, P.: *DEDALE, Système Expert pour dépannage de circuits électroniques analogiques.* (To be published.) Internal Research Report Fxxx, IBM Paris Scientific Center. 1985.

[4] Fargues, J.: *Contribution à l'étude théorique du raisonnement.* Doctoral thesis, Paris 6 University. 1983..

[5] *Artificial Intelligence*, special volume about qualitative reasoning, vol. 24, no. 1–3, Dec. 1984.

APPENDIX I

```
C ( t1 , t3 )  IS-A     circuit
        WITH    size         : rectangle ( 0 , 0 , 280 , 190 )
        AND     test-points  : t1 ( 5 , 100 ) ; t3 ( 270 , 80 )
        AND     composition  : ST1 ( t4 , t5 ) ; ST2 ( t6 , t7 ) ;
                               ND1 ( t1 , t4 , t6 ) ; ND2 ( t3 , t5 , t7 ) .

ST1 ( t1 , t2 ) IS-A    block
        WITH    reference    : translate ( 13 , 53 )
        AND     pattern      : m ( t1 , R1 , t2 , R2 , Ll1 ) .

m ( t1 , ru , t2 , re , li)
        WITH    size         : rectangle ( 2 , 2 , 72 , 117 )
        AND     composition  : ru ( t1 , t11 ) ; re ( t11 , t2 ) ;
                               li ( t11 )
        AND     test-points  : t1 ( 2 , 47 ) ; t2 ( 67 , 47 ) .

ST2 ( t1 , t3 ) IS-A    block
        WITH    reference    : translate ( 90 , 55 )
        AND     size         : rectangle ( 0 , 0 , 210 , 145 )
        AND     test-points  : t1 ( 0 , 45 ) ; t3 ( 205 , 45 )
        AND     composition  : R3 ( t1 , t2 ) ; R4 ( t2 , t3 ) ;
                               Ll2 ( t2 )       .

ru ( t1 , t2 )  IS-A    atom
        WITH    technology   : thin-layer
        AND     pattern      : rr ( t2 , t1 )   .

rr ( t1 , t2 )
        WITH    size         : rectangle ( 0 , 42 , 32 , 48 )
        AND     test-points  : t1 ( 32 , 45 ) .

re ( t1 , t2 )  IS-A    atom
        WITH    size         : rectangle ( 34 , 42 , 65 , 48 ) .

R3 ( t1 , t3 )  IS-A    atom
        WITH    size         : rectangle ( 0 , 42 , 100 , 48 )
        AND     test-points  : t3 ( 102 , 45 ) .

R4 ( t1 , t2 )  IS-A    atom
        WITH    size         : rectangle ( 104 , 42 , 203 , 48 ) .

ND1 ( t1 , t2 , t3 )  IS-A  internal-node .
ND2 ( t1 , t2 , t3 )  IS-A  internal-node .

li ( t )    IS-A    link .
Ll2 ( t )   IS-A    link .

C ( n1 , n2 )   IS-A    resistor ( [n1 , n2 , 15] )
        WITH    sub-fonctions : B1 ( n1 , n1 , n3 , n4 ) ;
                                B2 ( n3 , n4 , n2 , n2 ) .

B1 ( n1 , n2 , n3 , n4 )  IS-A    function
        WITH    impedance   : ( 7.5 )
        AND     pattern     : b ( n1 , n2 , n3 , n4 , R1 , R3 ) .

B2 ( n1 , n2 , n3 , n4 )  IS-A    function
        WITH    impedance   : ( 7.5 )
        AND     pattern     : b ( n1 , n2 , n3 , n4 , R2 , R4 ) .

b ( n1 , n2 , n3 , n4 , r1 , r2)
        WITH    sub-fonctions : r1 ( n1 , n3 ) ;
                                r2 ( n2 , n4 ) .

r1 ( n1 , n2 )  IS-A    resistor ( [n1 , n2 , 30] ) .

r2 ( n1 , n2 )  IS-A    resistor ( [n1 , n2 , 10] ) .
```

Fig. 7 — Example of the description language defining the specific circuit to be debugged.

APPENDIX II

```
<rule>   :=   <rule-name> [ ( <arg> [ <arg> ]*) ]
                IF   <premises>
                THEN   <concl> [ AND <concl> ]* .

<arg>   :=   <prolog-variable>

<premises>   :=   <prem> [ AND <prem> ]*

<prem>   :=   <integer> AMONG <premises>
              OR <premises> [ OR <premises> ]*   |
              <premises> IMPLY <premises>   |
              NOT <fact>   |
              <elementary-prem>

<elementary-prem>   :=   <fact>   |   <frame-predicate>   |
                         EXEC <prolog_goal>

<fact>   :=   <prolog-atom>   |   <prolog-predicate>

<frame-predicate>   :=
      <obj> ; [ <attribute> ; [ <operator> ; ]] <values>

<obj>   :=   <object>   |   <prolog-variable>

<object>   :=   <prolog-atom>   |   <prolog-predicate>

<attribute>   :=   <prolog-atom>   |   <prolog-predicate>

<operator>   :=   <attribute>   |   <name-of-binary-operator>   |
                  <prolog-predicate>

<values>   :=   prolog list of <elem-value>   |
                <elem-value> [ & <elem-value> ]*   |
                <prolog-variable>

<elem-value>   :=   prolog list of <elem-value>   |
                    <object>   |
                    <prolog-atom>   |   NOT <prolog-atom>   |
                    <prolog-number>   |   <prolog-string>

<concl>   :=   <fact>   |   NOT <fact>   |
               <frame-action>   |   EXEC <prolog-goal>

<frame-action>   :=   <obj> ; [ <attribute> ; ] <val>

<val>   :=   <elem-value>   |   <prolog-variable>
```

Fig. 8 — Syntax of rules.

6

Logic-based tools for building expert and knowledge-based systems: successes and failures of transferring the technology

Peter Hammond, Logic Programming Group, Department of Computing, Imperial College, 180 Queen's Gate, London SW7 2BZ, UK

INTRODUCTION

In recent years, many expert system shells and tool kits for building knowledge based systems have become commercially available. Inevitably, the kinds of application tackled, the structure of knowledge bases and the successful completion of end-products have been very much influenced by the formalism at the heart of the particular software package employed. For the originator of the shell or tool-kit, then, there is the opportunity to transfer a particular approach to knowledge representation to an audience wider than the immediate research group or commercial software house where the tools are initially developed.

The first and some of the second generation expert system tools have been around long enough now for an appraisal of both the tools themselves and their success or failure in the technology transfer of their underlying knowledge-based methods. This paper records some early findings of a survey of users of one particular toolkit, the augmented Prolog software, **apes**. The first versions of the **apes** software were originally produced in about 1981 on a project based at the Logic Programming Group at Imperial College and funded by the British Science and Engineering Research Council. Since the summer of 1984, commercial versions of the software have been marketed (Hammond and Sergot, 1985) under licence from the British Technological Group (to whom the **apes** intellectual property rights belong under the original funding agreement).

The fact that **apes** is implemented in Prolog, as opposed to Lisp or even some conventional programming language such as Pascal, is highly significant. It is Prolog's roots in logic programming that are very much emphasised in **apes** whether it is being used for a domain-specific application by an end-user, by a programmer or knowledge engineer (these two terms are considered equivalent in this paper) to develop an application or by a tool configurer/builder in altering the standard configuration of **apes** or in reconfiguring a subset of the **apes** facilities for an application other than expert systems.

Although **apes** can execute **any** Prolog program, non-declarative features included, correct interactive behaviour and explanations are guaranteed only for those programs which possess an obvious declarative reading. This appears at first, to some **apes** users, to be a restriction. Hopefully it is seen eventually as a benefit, especially when the importance of intelligibility and flexibility of use are considered. Moreover, the deductive model of problem solving (Kowalski, 1979) that is inherent and which is advocated in logic programming is a clear one for all users to understand and should be part of the armoury employed by programmers and knowledge engineers alike when constructing knowledge bases.

This common usage in **apes** of a deductive model by end-user, programmer and tool builder alike is further matched by their use of rule-based programming methods. As a rule-based programming language for the rapid prototyping of software, Prolog is second to none — although, as is illustrated in a later example, the implication of naive specifications can be both tempting and problematic.

Many of the present generation of knowledge engineers are now keen to develop their own software tools, having worked, sometimes unhappily, with "brought-in" expert system shells and tools. Typically, they would like to collect together an array of software tools with features that they have found useful or, at the very least, interesting. As tool configurers and builders, they need access to the primitives built into the shells and tools they currently use in order to replace and augment them as required. In more recent versions of **apes**, an effort has been made to give such lower level access to the underlying features and facilities, whether they control interaction, explanation or the problem solving process itself.

It seems to be unavoidable that end-users become knowledge engineers and programmers, and that knowledge engineers become tool builders and configurers. One way of assisting this quite natural development is to provide an homogeneous but configurable software environment in which to develop knowledge based systems at these different levels. The use of Prolog in **apes** as declarative rule-based programming language provides just such an environment.

THE APES USERS

Until recently, the vast majority of **apes** users have been programmers and knowledge engineers. There are few end-users of purpose-built applications (known to the author) and a small number of tool builders/configurers. There are more than five hundred sites where **apes** is being (or has been) used in a wide variety of application domains. A survey of users has recently been stated in which, initially, contact is made by telephone, but if it seems to be mutually beneficial (and it usually is), a site visit is made. Such direct contact enables users to suggest improvements in the software. It also gives the **apes** designer the opportunity to inspect knowledge base and program design at close quarters while at the same time reducing problems of confidentiality which discourage many users (mostly commercial and industrial) from allowing program listings to be made available away from the development site.

Later in this paper, examples are given of cases where misuse of the **apes** features has occured because of miconceptions about the **apes** software itself, or because of clashes between training and significant expertise in more traditional computer languages and lack of experience of declarative methods.

Table 1 summarises the distribution of the sites where **apes** is being used both geographically and according to a fairly loosely defined categorisation.

Table 1 — Numbers of apes sites by location and (approximate) area of application (March 1986).

	USA and Canada	UK	Europe (not UK)	Elsewhere	Total
Commerical	2	7	0	1	10
Computing	46	25	14	1	86
Education	58	67	32	20	177
Engineering	32	18	6	3	59
Government	19	14	0	0	33
Media	0	2	0	0	2
Medicine	1	7	2	0	10
Service	2	5	0	0	7
Unknown	54	13	17	7	91
Total	214	158	71	32	475

These figures, of course, do not give any accurate reflection of the full spectrum of users of expert systems software in general and are very much influenced by the availability and advertising of **apes** itself in the locality — the main distributors and dealers of the software being in Australia, Austria, Belguim, Canada, Finland, Holland, Japan, Norway, UK and USA. However, the figures for computing, education, engineering and government usage are significant. In particular it is interesting to look more closely (Table 2) at those for engineering and government.

The legislatory/regulatory applications in Table 2 are actually more widespread than the government figures suggest. Rule-based representations of pension regulations, building regulations, internal company

Table 2 — A more detailed breakdown of engineering and government sites.

Engineering		Government	
Aerospace	3	Atomic research	1
Construction	1	Building research	1
Electrical/electronic	18	Chemical analysis	1
Nuclear	3	Chemical defence	1
Petrochemical	9	Communications	2
Process control	2	Computer advisory	1
Unidentifiable	18	Crime detection	1
		Environment	3
		Legislative/regulatory	2
		Military	7
		Taxation	2
		Training	4
		Unidentifiable	7

regulations are other examples of this kind that are known to be under investigation and it is very likely that they form one of the most popular application domains.

Identifying what users are interested in and in which areas they have actually built systems, even prototypes, can only be accurately assessed by fairly close contact. Systems that appear in correspondence or in conversation to be of significant proportions may turn out to be quite insubstantial after investigation by an experienced logic programmer. This can largely be put down to the inexperience of the programmer with declarative programming or even a lack of awareness of the power and generality of a language like Prolog. In particular, the number of rules in a system is an inadequate estimate of complexity since by exploiting the power of Prolog programming techniques, the knowledge base can often be significantly reduced in size.

Table 3 gives some idea of the application domains identified so far after contacting less than 10% of **apes** users. Over the next six months or so, it is hoped that a substantial proportion of the remaining users will be contacted. Where they exist, references to relevant publications are given. There is an expected paucity of references to published descriptions of commercial or industrial projects. The majority of the available references are concerned with small to medium sized projects based in academic institutions.

LOGIC PROGRAMMING

The rest of this paper is more concerned with the application of logic programming techniques rather than the formalism and so only a brief description of logic programming itself is necessary.

Horn clause logic programs consist of a set of rules and assertions/facts

(Kowalski, 1982). An assertion expresses an atomic relationship among individuals. For example,

> aspirin aggravates peptic-ulcer
> lomotil aggravates impaired-liver-function
>
> Peter complains-of pain
> Peter complains-of inflammation
> Marek complains-of inflammation
> Peter suffers-from peptic-ulcer
> ...

define the relationships called **suffers-from**, **complains-of**, **suppresses** and **aggravates** amongst the individuals **Peter**, **Marek**, **aspirin**, **diarrhoea**, **imparied-liver-function**, **inflammation**, **lomotil**, **pain** and **peptic-ulcer**.

A **rule** in a logic program has the form

> A if B and C and ... and D

of a single conclusion A and one or more conditions B, C, ..., D. Conclusions and conditions, like assertions, express atomic relationships. Besides **constants** (such as "pain" and "aspirin"), rules and assertions can contain **variables** (denoted throughout by a leading underscore) which stand for **any** individual. A simple example of a rule is

> _drug may-harm _person if
>   _drug aggravates _condition and
>   _person suffers-from _condition

A logic program is invoked by posing a query such as

> **confirm(aspirin may-harm Peter)**

which would cause the answer **Yes** to be returned, because **aspirin may-harm Peter** follows logically from the rule and the assertions.

An important extension to Horn clauses is that of negation. In negation-by-failure (Clark 1978) a negated condition **not (P)** is considered to hold if the unnegated condition **P** fails to hold. A second drugs rule makes use of a negated condition

> _person should-take _drug if
>   _person complains-of _symptom and
>   _drug suppresses _symptom and
>   not _drug may-harm _person

and the query

> **find(_drug : Peter should-take _drug)**

returns the answer

> _drug = indomethacin

because **indomethacin may-harm Peter** cannot be deduced from the first rule and hence **Peter should-take indomethacin** can be duduced from the second rule.

Queries, of course, can involve conjuctions of relationships such as

> find(_drug: Marek should-take _drug and _drug may-harm Peter)

which, once again as a logical deduction, would return the answer

> _drug = aspirin.

To summarise, then, an answer to a query is an abbreviation for a sentence which can be deduced as a logical consequence of the program. This deductive model of problem solving enables us to understand logic programs **declaratively** without reference to the computer. However, if logic programs are to be executed on computers, then another interpretation, the **procedural** interpretation, is required.

Rules of the form

> A if B and C and ... and D

can be interpreted (and implemented) as **procedures** which reduce problems (goals) of the form **A** to subproblems (subgoals) of the form **B, C, ... and D**. Assertions do not introduce any subgoals and so terminate one path of the problem solving process. Thus, the problem of finding a drug that Peter should-take is reduced to the subgoals of finding symptoms that he complains-of and drugs that suitably suppress those symptoms but do not actually harm him.

The application of rules and assertions to solve a selected problem involves the use of the **resolution rule** of inference (Robinson 1965), and **unification** is employed to match the conditions in the query (or the conditions of a rule) to the form of rules or assertions in the program or knowledge base. The selection of rules and the order of the evaluation of condition is immaterial to the deductive process, their specification is determined when the procedural interpretation is mapped to a particular computer architecture (Kowalski, 1982).

PROLOG

Prolog is a logic programming language in which the selection of rules and the order of the evaluation of conditions is sequntial, in the order in which rules are defined and in the order in which conditions appear in rules and queries. When a procedure call fails, Prolog backtracks to the next untried procedure. On the other hand, control continues to evaluate the remaining conditions when the current procedure call is successful. Thus, Prolog is just one possible mapping of the procedural interpretation into computers architectures that involve sequential processing.

A raw Prolog system does not interact with a user except when accepting and supplying answers to queries. Moreover, the set of rules and facts employed in a successful procedure call and the variable substitutions generated by the corresponding unification (except for those relevant to the query) are not recorded and so explanations of program behaviour are not immediately available for the user's inspection.

THE APES SYSTEM

The **apes** software is itself implemented as a collection of Prolog modules. Amongst other things, **apes** augments the underlying Prolog with two main features.

(a) Declarative interactive behaviour

Some relationships in a logic program may be unspecified when a query is posed. A Prolog system will simply generate a system error indicating the absence of defining clauses (rules or facts). **apes** incorporates an implementation of the Query-the-User model which handles interaction declaratively (Sergot, 1983) and requests information from the user about any undefined relationship that has been generated as a subproblem in the solution of a query. The programmer is able to avoid the otherwise inhibiting use of direct calls, within knowledge base rules, to Prolog primitives for handling input and output.

(b) Explanations of program behaviour

As in usual in almost all expert system tools, **apes** provides explanations of how questions to the user have arisen from the consideration of rules in the knowledge base, how queries have been solved and why some queries cannot be solved. In a logic-based system such as **apes**, an explanation of an answer to a query is an annotated proof of the deduction from the program of the sentence representing the answer.

Even with the basic facilities, it is possible to build rudimentary logic-based expert and knowledge based systems of some interest. **apes** posseses many other features, most of which are ignored in this paper, for implementing knowledge bases of considerable size and complexity.

RAPID PROTOTYPING AND PROGRAM DEVELOPMENT

Significantly, users of Prolog generally praise its support for the rapid prototyping of software. In many of the applications mentioned earlier, **apes** interactive facilities (described in more detail later) have been employed in addition to Prolog's inherent protyping capability. By developing programs top-down and running queries within **apes**, the programmer can simulate the desired behaviour of lower level primitives that are yet to be defined, since **apes** is forced to ask about any such undefined relationships. Thus, the programmer can concentrate on establishing correct higher level behaviour without worrying about lower level programming detail.

However, even when program behaviour appears to be satisfactory, the initial specification of the expert's knowledge frequently proves to be inadequate and refinement of the knowledge base is unavoidable. In some cases, rules represent an inaccurate picture of the domain knowledge. Subtle changes may be needed to extend or reduce the scope of rules, or it may be that extra parameters are required to account for some previously neglected aspect. The drugs example is a suitable vehicle for illustrating this evolutionary nature of developing a knowledge base. A simple extension is demonstrated below; a more complete version of the program formed part of a knowledge base translated from EMYCIN to **apes** in 1983 (Alvey & Hammond, 1983).

As they stand, the collection of drugs rules and facts could be used as a very rough and ready pharmacopoeia to prescribe a very limited number of drugs for an equally limited number of symptoms. One immediate drawback is its insensitivity to the efficacy of drugs according to their route of administration and the suitability of the associated format (tablets, syrup, injection etc.). This can be partly remedied by redefining the top-level rule to take account of the extra parameter.

```
should-take(_person _drug _format) if
    _person complains-of _symptom and
    treatment(_drug _format _person _symptom) and
    not suitable (_drug _format _person).
```

Two ways for assessing the unsuitability of a drug treatment can now be catered for

```
unsuitable(_drug _format _person) if
    _person suffers-from _condition and
    _drug aggravates _condition

unsuitable(_drug _format _person) if
    _person complains-of inability-to-swallow and
    not _format given-as liquid.
```

Instead of the simple association between a drug and a symptom that it suppresses, the treatment can now be more sensitive:

```
treatment(_drug _format _person _symptom) if
```

>   needs-therapy(_person _symptom _therapy) and
>   provides-therapy(_drug _route _therapy) and
>   _drug available-as _format
>
> needs-therapy(_person pain weak-analgesic) if
>   _person complains-of pain and
>   severity(_person pain mild)
>
> needs-therapy(_person pain psychoactive) if
>   _person complains-of pain and
>   severity(_person pain overwhelming)
>
> needs-therapy(_person pain anti-inflammatory) if
>   _person suffers-from painful-bone-metastases
>
> needs-therapy(_person vomiting anti-emetic)
> needs-therapy(_person nausea anti-emetic).

The drug data now becomes

> provides-therapy(aspirin oral weak-analgesic)
> provides-therapy(pethidine oral medium-analgesic)
> provides-therapy(pethidine intra-muscular strong-analgesic)
> provides-therapy(cyclizine oral anti-emetic)
>
> tablets format-for-route oral
> soluble format-for-route oral
> capsules format-for-route oral
> lozenges format-for-route oral
> injection format-for-route intra-muscular
> suppositories format-for-route rectal
>
> aspirin available-as tablets
> aspirin available-as soluble
> morphine available-as injection
> morphine available-as tablets
>
> syrup given-as liquid
> suspension given-as liquid
> soluble given-as liquid
> tablets given-as solid
> capsules given-as solid
> lozenges given-as solid.

So, by concentrating on one or more aspects of the prescribing process, and using the modifiability of the rules, a new more useful set can be generated. The need for such changes cannot always be anticipated so a formal specification stage, followed by implementation and testing is unsuitable. Instead, the rules themselves must act wherever possible as both specification and program (Kowalski, 1984).

SPECIFICATIONS AND EFFICIENCY

In all applications, it is important to review carefully any programs that have been developed from informal specifications and implemented directly in Prolog. Some application domains require high levels of program performance, even on relatively low-powered micro-computers. Just such as a situation arose from an application of **apes** in the development of programs to aid the solution of secondary school level chemistry problems (Bateman 1985). This example illustrates features that are likely to be common to many other application domains and so is worth highlighting.

The selection of techniques suitable for separating various mixtures of chemicals into their consituent parts is one of the problems considered. Such a problem requires characterisations of mixtures in terms of the solids and liquids they contain as well as in terms of their solubility and miscibility. A natural informal specification (somewhat simplified for this illustration) of part of the classification might be as follows:

> a mixture is a SOLID-MIX if every component of the mixture is a solid
>
> a mixture is a LIQUID-MIX if every component of the mixture is a liquid
>
> a mixture is a SOLID-LIQUID-MIX if every component of the mixture is either solid or liquid
>
> a mixture is a SUSPENSION if it is a SOLID-LIQUID-MIX and every liquid component is miscible and every solid component is insoluble.
>
> a mixture is a SOLUTION-MIX if it is a SOLID-LIQUID-MIX and every liquid is miscible and every solid is soluble

The selection of suitable separation procedures is described by statements such as

> dissolution process 14 is suitable for a SOLUTION-MIX
>
> crystallization process 13 is suitable for a SUSPENSION
>
> fractional-crystallization process 2 is suitable for a SOLUTION-SUSPENSION mix.

The classificatory rules can be expressed as logic programs making use of the high level **forall** primitive:

> _mixture is-a SOLID-MIX if
>> (forall _component part-of _mixture then _component is solid)
>
> _mixture is-a LIQUID-MIX if
>> (forall _component part-of _mixture then _component is liquid)

_mixture is-a SOLID-LIQUID-MIX if
  _component1 part-of _mixture and _component is solid and
  _component2 part-of _mixture and _component2 is liquid
_mixture is-a SUSPENSION if
  _mixture is-a SOLID-LIQUID-MIX and
  (forall _component part-of mixture
    and _component is liquid
    then _component is-miscible) and
  (forall _component part-of _mixture
    and _component is solid
    then _component is-insoluble)
_mixture is-a SOLUTION if
  _mixture is-a SOLID-LIQUID-MIX and
  (forall _component part-of _mixture
    and _component is liquid
    then _component is-miscible) and
  (forall _component part-of _mixture
    and _component is solid
    then _component is-soluble)

The transfer from informal specification to program, then, was made particularly straightforward by the use of the **forall** and was indeed a very natural represenation to use. However, it is not surprising that these rules appeared to the programmer to be decidedly sluggish in performance, considering the amount of recomputation that is carred out. In order to remove the obvious inefficiencies, it is important to co-routine the analysis so that miscibility and solubility, let alone liquidity and solidity, are established together wherever possible. So, although, during the prototype stage, the iterative nature of the **forall** structure was very useful to a naive user, when it came to run-time efficiency, the programs performed unsatisfactorily.

There are various ways of rewriting these rules and improving their efficiency, but it is likely that none will be so close in character to the original informal specification as the use of the **forall** construct. A partial solution, not giving an exhaustive classification, would be a program of the following form which first separates the individual components symbolically, something which only appears to be important when a more detailed inspection of the separation selection is made.

classify(_solids _liquids _class) holds when the collection of components in a mixture represented by the list of solids **_solids** and list of liquids **_liquids** is of the type **_class**. Notation used throughout is **(_head| _tail)** for a list whose first element is **_head** and whose remainder is **_tail**. The empty list is denoted by ().

  if there are no liquids and at least one solid, then the mixture is a
  SOLID-MIX

classify((_solid| _rest-of-solids) () SOLID-MIX)

if there are no solids and at least one liquid, then the mixture is a LIQUID-MIX

classify(() (_liquid| _rest-of-liquids) LIQUID-MIX)

otherwise

classify((_solid|rest-of-solids) (_liquid|_rest-of-liquids) SOLID-LIQUID-MIX)

analyse(_mixture _solids _liquids) holds whenever the mixture **_mixture**, the list of components, is composed of liquids **_liquids** and solids **_solids**. A first obvious definition of **analyse** is

analyse (() () ())

analyse ((_component|_rest-of-mixture) (_component|_solids) _liquids) if _component is solid and
analyse (_rest-of-mixture _solids _liquids)

analyse ((_component|_rest-of-mixture) _solids (_component|_liquids)) if _component is liquid and
analyse (_rest-of-mixture _solids _liquids)

Even these two rules for **analyse** have a redundant inefficiency in testing the type of each individual component twice. They can be combined into the more efficient program

analyse (() () ())

analyse ((_component|_rest-of-mixture) _solids _liquids) if _component is _type and
further-analyse(_type _component _rest-of-mixture _solids _liquids)

further-analyse(solid _comp _mixture (_comp|_solids) _liquids) if analyse(_mixture _solids _liquids)

further-analyse(liquid _comp _mixture _solids(_comp|_liquids)) if analyse(_mixture _solids _liquids)

This improvement in efficiency unfortunately brings with it a substantial reduction in intelligibility because the program is now much removed from the original informal specification. The explanations of how the analysis has been carried out will be equally lacking in clarity.

QUERY-THE-USER

The Query-the-User component of **apes** was referred to earlier and here a more detailed description is given in preparation for a later illustrative example.

In cases where the user poses queries which may involve (directly in the query conjunction or indirectly in the reduction to sub-problem solving) relations still to be defined, the Query-the-User component of **apes**, in its simplest implementation, immediately puts a question to the user in an attempt to retrieve the information that is required. It is as if the user is viewed as an extension to the knowledge base and a source of other missing assertions.

If, for example, the assertions for the relations for **complains-of** and **suffers-from** where removed from the drugs program, the query

find(_drug : Peter should-take _drug)

would generate a short dialogue such as the following (user's input is shown **bold** and the actual answer process has been simplified so as not to clutter the illustration with unnecessary detail)

   Which _A : Peter complains-of _A ?
    ==> **pain**   the user's answers
    ==> **diarrhoea**
    ==> **inflammation**
    ==> **end**   indicating there are no more
   Is it true that Peter suffers-from peptic-ulcer ?
    ==> **yes**
   Is it true that Peter suffers-from impaired-liver-function ?
    ==> **no**
   Answer to query : _drug = indomethacin **more** (asking for more
                  answers)
   Answer to query : _drug = lomotil

The questions **apes** has generated are default forms used in the absense of any overriding instructions provided by the programmer in the form of alternative and preferred question templates. The assertions

   is-template(suffers-from (_person peptic-ulcer)
    (Does _person have problems such as peptic ulcer ?))
   is-template(suffers-from (_person imparied-liver-function)
    (Has _person had liver trouble recently ?))
   which-template(complains-of (_person _symptom) (_symptom)
    (Give the symptoms which _person
    is currently complaining of))

would alter the dialogue of

   find(_drug : Peter should-take _drug)

Give the symptoms which Peter is currently complaining of
==> **pain**
==> **diarrhoae**
==> **inflammation**
==> **end**

Does Peter have problems such as peptic ulcer ?
==> **yes**

Has Peter had liver trouble recently ?
==> **no**

Anser to query : _drug = indomethacin

etc.

There is no need to restrict these question descriptions to assertions. Much more flexible question forms can be defined:

the two rules

    is-template(taken-medicine(_person) (Has _person taken his medicine)) if _person is-male

    is-template(taken-medicine(_person) (Has _person taken her medicine)) if _person is-female

would cause questions such as

    Has Peter taken his medicine?

    Has Mary taken her medicine?

assuming, of course that

    Peter is-male   and   Mary is-female

could be deduced from the knowledge base.

An alternative and more compact way to define these two question templates is

    is-template(taken-medicine(_person) (Has _person take _A medicine)) if
    _gender is-sex-of _person and
    _A is-possessive-pronoun-for-sex _gender

his is-possessive-pronoun-for-sex male

her is-possessive-pronoun-for-sex female

This example illustrates how the same programming power that Prolog offers the programmer in describing the expert's knowledge is also available for affecting how questions which arise in query invocations appear to the user. When used with imagination, these features can give great flexibility and variation during the **apes**-user interaction.

MISUNDERSTANDING THE QUERY-THE-USER AND DEDUCTIVE MODELS

From the close inspection of a number of knowledge bases built using **apes**, it is obvious (and understandable) that some programmers find it difficult to "shake off" their roots in traditional programming. The idea that programs should generally be structured according to the three phases **input data**, **process data**, **produce output** is apparent all to often for it to be ignored. This pre-occupation with program control is even suggested in those Prolog texts which emphasise the **procedural** interpretation of Prolog programs over and above the **declarative** one. The most common error is easily demonstrated with reference to the drugs example.

The rules

finished processing if
  get imput and
  processed input and
  output reported
get input if
  Peter complains-of _drug and
  Peter suffers-from _condition
processed input if
  Peter should-take _drug and
  add(_drug is-drug-for Peter)
output reported if
  _drug is-drug-for Peter and
  print(Peter should-take _drug)

together with the **apes** query

confirm (processing finished)

would generate, given the relevant question definitions, a dialogue of the form

Give the symptoms of which Peter is complaining?
==> **pain**
==> **inflammation**
==> **end**

From which medical conditions is Peter suffering?
==> **peptic-ulcer**
==> **end**

Peter should-take indomethacin

⋮

On the surface, the interaction is reasonably acceptable. Underlying it, though, is a highly procedural view of the process of selecting a suitable medication. This example illustrates a view that programming is a matter of executing procedures so that all relevant data is at hand before the main computation takes place and the "final report" is generated as output. Such programs are marred by

(i) a pre-occupation with total control of the computation;
(ii) the anticipation of obtaining data (that supplied by the user) for later processing even though it may not be relevant;
(iii) the assignment, unnecessarily (using the built in non-logical Prolog primitive **add** for dynamic additions to the program) of values to be used in an output stage;
(iv) a total misunderstanding that the interactive and reporting behaviour is provided in any case by **apes** built-in features.

DYNAMIC ALTERATION OF THE KNOWLEDGE BASE

Configuration problems appear to be a popular type of application, most frequently in electronic and engineering applications. One **apes** user has developed a system for a British engineering company for configuring valves systems. The expertise that has been embedded within the system enables some customer specifications to be evaluated in 20 minutes instead of the half day normally required by the less experienced staff [Mobbs, 1985].

On approach to configuration problems that most **apes** users experiment with involves the dynamic alteration of the knowledge base as the evaluation of a particular problem develops. Apart from the fact that **apes** would be unable to explain such programs, there is the additional difficulty of establishing that the program actually gives a complete or correct solution to the configuration problem. The application domain considered below is similar to a configuration problem considered by another **apes** programmer.

The problem of protecting components against various kinds of hazards is widespread in industrial and civilian uses of electronic devices. The hazards can range from power surges in electrical supply to X-ray and heat radiation in hospital radiography departments. The materials available for the construction of electronic components have varying degrees of susceptibility to different hazards and protection can often be provided locally to an individual component, globally to the entire device or even to some large substructure. The protective devices may themselves be composed of different materials.

From a conventional programming point of view, an obvious model to employ is one of generating a database of assertions describing the configuration of protective devices with respect to components, materials and hazards, each assertion being of the form:

protection(component-1 material-3 hazard-10 protection-8).

For the moment, assume that once each component has been considered for a particular hazard there is no further alteration to the corresponding assertion as a result of the subsequent addition of new assertions describing other risks and components. This simplified version of the problem has a straightforward solution and requires an iteration through the various components and hazards selecting suitable materials and a mechanism for their protection.

The assertion generated for each cycle of the iteration is in fact output that becomes part of the knowledge base. The notation **add(C)**, where C is a structured term, is used to represent the dynamic addition of the assertion C to the program.

A rule describing the selection of a suitable protection might look as follows:

needs-protection(_comp _hazard _material _protection) if
_material suitable-material-for _comp and
susceptibility(_material _comp _hazard _susc) and
estimated-risk(_comp _hazard _risk) and
_susc LESS _risk and
protection-device(_susc _risk _protection).

In addition, the knowledge base requires definitions of suitable materials, their susceptibilities to each risk and rules for calculating the risk given the existing usage of protective devices. The generation of new assertions is carried out in the higher level rule:

_comp is-protected-against ()
_comp is-protected-against (_hazard|_rest-hazards) if
    needs-protection(_comp _hazard _material _protection) and
    add(protection(_comp _hazard _material _protection)) and
    _comp is-protected-against _rest-hazards.

Once the new assertion for **protection** has been added to the knowledge base, it can be used in subsequent calls to **is-protected-against** and hence can influence the consideration of the remaining hazards. For example, a global protective device may be the only way of shielding one component against a particular risk. Once a clause describing the assignment of that protective device has been asserted, later calculations of estimated risks are necessarily affected.

Now the use of the **add** primitive may be clearly understood when considered in isolation, with respect to the individual rule that invokes it. But its global effect on the knowledge base and on a particular problem solution may not be very clear. Therefore, the logical correctness of any solution must be in some doubt. However, as is explained in [Hogger, 1984, p. 119], it is sometimes possible to explain the execution of a program containing calls to **add** in terms of a closely related program which makes no such use of dynamic assertion. For example, suppose the call to **add** is removed but its arguments left behind in the definition of **is-protected-against**. Suppose also that the assertions generated in a particular problem solution are also added to the rest of the original knowledge base. The definition

```
_comp is-protected-against ()
_comp is-protected-against (_hazard|_rest-hazards) if
        needs-protection(_comp _hazard _material _protection)
        and
        protection(_comp _hazard _material _protection) and
        _comp is-protected-against _rest-hazards
```

is perfectly acceptable from a logical point of view since any particular solution is consistent with the rest of the knowledge base.

There may not always be an obvious, "clean", program to refer to with respect to an algorithm which makes use of dynamic assertion. The situation is even worse if the program makes use of dynamic removal of clauses from the knowledge base. As Hogger further states, the inclusion of negated conditions in such **add** and **delete**-ridden programs compounds the difficulty of establishing the logical correctness of solutions since "dynamic modification of the program alters what is **non-provable**".

Unfortunately, the next stage in the development of the configuration problem considered above would be to include some dynamic deletion of instances of the **protection** relation. These deletions do not arise because of any backtracking since the problem is largely deterministic in character. They are required because later assignment of protection can radically alter earlier selections of protective devices. For, if, as has been indicated earlier, some kind of global protection device is employed to shield a particular component then it makes sense to alter those instances of **protection** which may now be unnecessary or at least may require modification.

An alternative solution is to abandon the explicit database model altogether and manipulate instead a single term which captures the assignment or configuration of materials and protection to each component-hazard pair. The new definition of **is-protected** might be:

> is-protected(() _comp **_configuration _configuration**)
> is-protected((_haz|_rest) _comp **_old-config _new-config**) if
>   needs-protection(_comp _haz _mat **_old-config**
>   _protection) and
>   updates(**_old-config** _protection **_temp-config**) and
>   is-protected(_rest _comp **_temp-config _new-config**)

and the new definition for **needs-protection** might be

> needs-protection(_comp _haz _mat **_old-config** _protection) if
>   _mat suitable-material-for _comp and
>   susceptibility(_mat _comp _hazard _susc) and
>   estimated-risk(_comp _hazard **_old-config** _risk) and
>   _susc LESS _risk and
>   protection-device(_susc _risk _protection).

The definition of **estimated-risk** must now take into account the protection of those components already considered by retrieving the relevant information from the configuration term rather than from the **protection** assertions. The assimilation of a newly generated protection into the existing configuration is handled in the relation **updates**. Although these new definitions constitute a much "cleaner" program, there are some drawbacks. Firstly, the term describing the current state of the configuration is now taking up considerable stack memory during computation and this in itself can cause resource problems. Secondly, to assimilate the new protection it is necessary to search the term representing the configuration in a recursive manner and hence lose the exploitations of the inbuilt facilities for rapid data access to individual clause structures.

CONCLUSIONS

This paper has provided an opportunity to report on data currently being collected on the number and distribution of **apes** users. Application domains continue to diversify rapidly as the expert systems phenomenon (some say hysteria) spreads throughout all disciplines. It is still likely that **apes** users have more interest in building expert systems than in exploiting its logic programming features. It is obvious, but worth noting, that the underpinning knowledge based technologies will survive and continue to contribute to software design after expert systems have become a thing of the past.

Early indications from the survey of **apes** users that the deductive problem solving nature of logic programs as well as the interactive features of **apes** can easily be ignored if there is a pre-occupation with conventional control of Prolog programs. Although the rapid prototyping of knowledge bases is encouraged by a system such as **apes**, it is quite easy for inexperienced programmers to succumb to these high level features and develop programs that are correct but inefficient. Provided the programmer is aware of the limitations of such programs, and does not assume that **apes** or the underlying Prolog is necessarily the source of the inefficiency, then steps can be taken to improve performance.

The homogeneous environment that **apes** provides for end-user, programmer, knowledge-engineer and tool builder should encourage each to move on from one role to the next in the development of expert system applications as well as expert systems tools. However, given that there are signs that some users are less conscious of the declarative approach to building logic-based systems, caution should be applied if users are to experiment with new control strategies and the accompanying complexity of explanations and interactive behaviour.

On a more practical not, it is obvious to the author that academic researchers in logic programming, and the author is equally culpable, are failing to convince industrial users that Prolog and related software systems, such as **apes**, can be used, in the deductive rule-based style emphasised here, to build, or at least prototype, significant software applications. Many apes users complain that the logic programming literature is lacking in illustrative examples of sufficient complexity or relevance to their area of expertise. The examples contained in this paper are an attempt to make logic programming techniques more attractive and relevant to a wider audience.

Of course, the apparent lack of suitable tutorial material would be eradicated if expert system builders using Prolog advertised their work more widely. Perhaps, this also identifies a gap in the application of logic based methods that could be best filled by individuals with a foot in both the acedemic and industrial sectors. Some kind of "buffer" or "clearing house" is essential if the problems and applications arising outside academia are to be satisfactorily matched with the research activity that is generating problem solving methodologies within.

REFERENCES

Allwood, R. J., Stewart, D. J., Hinde, C., Negus, B., Report on Expert System Shells Evaluation for Construction Industry Applications, Loughborough University of Technology, Loughborough, U.K.

Alvey, P., Hammond, P., A Comparison of EMYCIN and **apes** for a Medical Application, Research Report, Department of Computing, Imperial College.

Bateman, D., private communication.

Blythe, H., Computerised Patient Interviewing Using micro-PROLOG and **apes**, M.Sc. Report (1985), Department of Computing, Imperial College.

Carey, S., An Expert System for Enhanced Diabetes Care, Artificial Intelligence for Society Brighton Polytechnic, July 1984.

Chan, D., A Logic-Based Legal Expert System, M.Sc Report (1984), Department of Computing, Imperial College.

Coady, W. F., Automated Link Analysis, Artificial Based Tool for Investigators, *Police Chief Magazine*, September 1985, USA.

Hamilton, I., Building Foundation Design, M.A. Dissertation.

Hammond, P., Representation of DHSS Regulations as a Logic Program, Proceedings of BCS Expert Systems 1983, Cambridge, and in *Models for Decision Making: Mathematical Progamming, Decision Analysis, Expert Systems* (G. Mitra, editor), North-Holland.

Hammond, P., Sergot, M. J., **apes** documentation, 1985, Logic Based Systems Ltd., 40 Beaumont Avenue, Richmond, Surrey, TW9 2HE, England.

Kowalski, R. A. (1979), *Logic for Problem Solving.* North Holland. Elsevier. New York.

Kowalski, R. A., (1982), Logic Programming, Invited paper at IFIP'83.

Kowalski, R. A., The relation between logic programming and logic specification, *Phil. Trans. R. Soc. Lond. A* **312**, 345–361.

Lehner, P. E., Barth, S. W., Expert Systems on Micro-Computers, PAR Technology Corporation, USA.

Lowes, D., Assistance to Industry: A Logical Approach, M.Sc. Report (1984), Department of Computing, Imperial College.

Mobbs, C., "Why users ignore Expert Advice" (J. Lennox), *PC Business World*, September 24th 1985.

Nelder, J. A., Wolstenholme, D. E., A Front End for GLIM, Proceedings of Expert Systems in Statistics Workshop, Aachen, December 1985.

Pearce, J., A Housing Benefit Expert System using APES, M.Sc. Report (1984), Department of Computing, Imperial College.

Pearse, R., Rosenbaum, M., Hammond, P., The Evaluation of Proposed Road Corridors by the use of an Expert System, Proceedings of Conference on Applications of Artificial Intelligence in Engineering, April 1986, Southampton, U.K.

Polonsky, M., Psychological Test Expert System, M.Sc. Report (1984), Department of Computing, Imperial College.

Robinson, J. A., (1965), A Machine-oriented Logic Based on the Resolution Principle. *JACM.* **12**. pp. 23–41.

Sergot, M. J., (1983), A Query-the-User facility for logic programming. In *Integrated Interactive Computer Systems* (P. Degano and E. Sandwall, eds.). North-Holland Publ., Amsterdam.

Sergot, M. J., Sadri, F., Kowalski, R. A., Kriwaczek, F., Hammond, P., Cory, H. T., British Nationality Act as a Logic Program (1985), Department of Computing, Imperial College.

Simons, G. L., Expert Systems and Micros, NCC Publications 1985, The National Computing Centre Ltd., Oxford Road, Manchester, M1 7ED, England.

Suphamongkhon, K., Towards and Expert System on Immigration Regulations, M.Sc. Report (1984), Department of Computing, Imperial College.

Tucherman, L., Furtado, A. L., Casanova, M. A., A Tool for Modular Database Design, Proceedings of VLDB 85, Stockholm.

Waterman, D. A. (1986) *A Guide to Expert Systems*, Addison-Wesley.

Yearbook of Law, Computers and Technology, Volume 2 (1986), Butterworths, London.

7

Model Guided Interpretation Based on Structurally Related Image Primitives

Gerd Maderlechner Siemens AG, D-8000 München 83, West Germany, **Eliane Egeli** and **Fernand Klein**, ETH, CH-8092 Zürich, Switzerland

1. INTRODUCTION

Documents including technical drawings or maps may be naturally segmented into objects and complementary background using standard methods of low level vision. These image components and their spatial relation must be extracted and represented. This relational description yields the topological and geometrical context for the interpretation with an application dependent model. The model is derived from the existing rules and standards of technical drawings.

Graph-theoretical methods [1, 2] have been successfully applied to the recognition of symbols in line drawings like flow charts or circuit diagrams. The reference symbols and the segmented line drawing are transformed into an undirected labelled graph. The symbol recognition is performed by subgraph isomorphism between symbol graph and image graph. Structured lines made of line primitives of different length and width and arranged in characteristic relative position (line textures) impede the recognition, because the spatial relations are not represented by the image graph and the model.

This chapter presents an approach, which overcomes these difficulties and allows improvements for the mentioned graph-theoretical methods, too.

2. EXTRACTION OF IMAGE PRIMITIVES AND THEIR SPATIAL RELATION

The description of the image primitives and their spatial relations is derived from a generalized (Euclidean) distance transform of the objects and the complement [3]. The Euclidean radius and additional attributes are determined for every object point. This flexible and redundant representation is compressed for applications to special object classes.

For line-like objects it is suitable to define the image primitives by their topological connectivity and to use the skeleton with the Euclidean radius as thickness information. The relational representation is performed by an attributed exoskeleton, which includes pointers to left and right neighbouring primitive and its distance (radius). The branches of the exoskeleton are borderlines between the "zones of influence" of the individual primitive elements (Fig. 1). The object skeleton and the exoskeleton (dual skeleton)

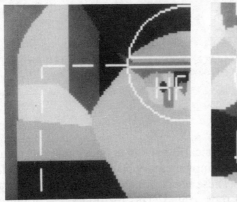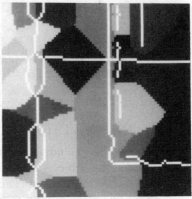

Fig. 1 — Object skeleton (white) and regions of influence (different grey levels) for dashed lines (left) and complex line structures (right).

and its additional attributes are combined to a triple of primitives (points, lines and nodes). Each primitive is identified by a key: pointnr, linenr, nodenr. The point primitive carries the coordinate (x,y) and distance (radius) information. The lines are determined by the keys of their startpoint and endpoint, because the points are sorted in a monotone way along the lines. Dual lines (of exoskeleton) have the additional information (leftline, rightline) of the neighbouring object primitives. Nodes result in vertices of the skeleton with three or four lines and contain the corresponding pointnr and the key of the incoming line and the next outgoing line in a definite circulation. These attributed primitives are used as terminal elements for an attributed grammar and are listed explicitly in the next section.

3. STRUCTURED IMAGE AND MODEL DESCRIPTION

The model description is based on an atttributed context-free grammar, which is generalized with additional semantic information. The structure of the symbol, symbol parts, primitives and their spatial relation are described by rules using predicate calculus. This approach avoids the sensitivity of standard syntactic methods on erors in the primitive classification, i.e. a sequence of primitives is checked on its compatibility with the syntactic and semantic rules of the given grammar. In our symbol recognition the rules control the selection of primitives which build a consistent sentence in the grammar [4]. Thus the grammar serves as a model for the symbol definitions and as a fast constraint in searching primitives belonging to the symbol [5].

This model description is implemented in Prolog, which offers an elegant formulation of the grammar and the semantic rules. The built-in inference machine using unification, resolution and backtracking proves the membership of a terminal element to the object model. In successful recognition the symbol description is instantiated with the corresponding primitive elements.

The model description is given by a 5-tuple

$$D = (T, N, S, PS, PA)$$

with
- S = Set of start symbols: symbol library
- T = set of terminals: image primitives
- N = set of nonterminals: symbols, symbol parts, join operations
- PS = structural and semantic rules, join rules
- PA = procedural attachment

The *terminal elements* T are composed of a syntactic part ‹name› and attribute values $\{a_1, \ldots, a_n\}$. The following attribute classes are defined:

‹Point›	{pointnr, x, y, radius}
‹Line›	{linenr, startpoint, endpoint}
‹Node›	{nodenr, linenr, pointnr, nextlinenr}
‹DPoint›	{dpointnr, x, y, radius}
‹DLine›	{dlinenr, startdpoint, enddpoint, leftline, rightline}
‹DNode›	{dnodenr, dlinenr, dpointnr, nextlinenr}

The *nonterminals* N are composed of a syntactic part ‹name›, a semantic part $\{p\}$ containing a representative (variable) of the class name and inherited attribute values $(f_1, \ldots, f_n)$.

The *rules* PS consist of:

— standard syntactic production rules
— constraint part (CONSTR) described by predicate calculus
— join part (JOP) linking corresponding elements
— assignment part (ASSIG) generating attribute values.

The standard syntactic part must be present in each rule, the remaining parts are optional.

The *procedural attachment* PA defines functions determining attributes of a structure, e.g. number of points on a line.

4. RECOGNITION OF A SIMPLE DASHED LINE

This line structure consists of regularly arranged short line elements. The characteristic neighbourhood relation is represented by a valley in the distribution of radius values on the dual line between the line elements (Fig. 2). The model description manages with only three nonterminals:

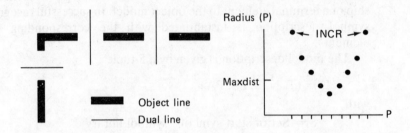

Fig. 2 — Neighbourhood relation and distribution of radius values for a dashed line. The parameters 'maxdist' and 'incr' are specific for this line structure.

'dotline' defines the properties of the line elements, 'joindot' describes the linking rule and 'dotsym' represents complete dashed line.

The complete model definition is shown in Fig. 3. The set T of terminal symbols is the same as above.

5. ADVANCED LINE STRUCTURES

The following line structures are investigated:

— C: dashed line parallel to continuous line
— B: three parallel continuous lines
— A: 'railway track'-line (railway symbol in maps)

The important structural relation *'parallelism'* is easily derived from the distribution of radius values on the dual lines. For type C short parallel elements are constructed which are linked by the corresponding continuous line or at vertices via the corresponding node. The line structure B is defined by longer parallel elements with a common line part. The 'railway track' A has a prominent structure defined by alternating short parallel elements with thick short line elements. The definition of each model uses one start symbol S and about three times more non-terminals N and rules in PS and PA than for the simple dashed line.

S = Dotsym
N = Dotsym, Dotline(dotlinerange = 7..18),
　　Joindot(incr = 4, maxdist = 7)
Ps =
1　‹Dotsym›　　{a.b}　　　　:: = ‹Dotline› {a} + ‹Dotline› {b}
　　JOP:　　　　‹Joindot›　{a, b} .
2　‹Dotsym›　　{[a].b.c}　　:: = ‹Dotsym› {[a].b} + ‹Dotline› {c}
　　JOP:　　　　‹Joindot›　{b, c} .
3　‹Dotsym›　　{a.b.[c]}　　:: = ‹Dotline› {a} + ‹Dotsym› {b.[c]}
　　JOP:　　　　‹Joindot›　{a, b} .
1　‹Dotline›　　{a}　　　　 :: = ‹Line› {a}
　　CONSTR:　nrpoint(a)　　in dotlinerange
1　‹Joindot›　　{a,b}　　　 :: = ‹Dline› {d11}
　　CONSTR:　element([dp1,dp2,dp3], ‹Dpoint›) &
　　　　　　　on([dp1,dp2,dp3], d11) & radius(dp1) < maxdist &
　　　　　　　key(dp2) = (key(dp1)+incr) & key(dp3) = (key(dp1)−incr) &
　　　　　　　radius(dp3) > radius(dp1) & radius(dp2) > radius(dp1)
　　ASSIG:　　(key(a) = left(d11) & key(b) = right(d11)) #
　　　　　　　(key(a) = right(d11) & key(b) = left(d11)) .

　　PA = (definition of following functions)
　nrpoint(1)　　　　　　: determines the number of points in line C.
　element(P,Dpoint)　　 : tests, if p belongs to 'Dpoint'.
　on(p,1)　　　　　　　: tests, if point p is on line 1
　key(p)
　radius(p)　　　　　　　extract the corresponding attribute
　left(1)　　　　　　　　values of the image primitives
　right(1)
　　Notation:　　+　　: means the join operator (JOP)
　　　　　　　　&,　 : logic AND resp. OR
　　　　　　　　a.b　: linked structure with parts a and b
　　　　　　　　[a].b : linked structure of same hierarchy
　　　　　　　　[a,b,c] : set of elements a,b,c

Fig. 3 — Model definition of dashed line (line structure D).

6. RESULTS AND DISCUSSION

All four line structures in Fig. 4 are successfully instantiated if the models are applied in a hierarchical order, beginning with the most stable structure A and ending with the weakest structure D (dashed line). The recognized structured lines are indicated by two arrows pointing to their endpoints. The display only shows those primitive elements, which belong to the instantiated symbols.

Type C is recognized although the dashed line touches the parallel continuous line. This is an example of the flexibility of our model. The error in the drawing (melting of lines) is compensated by a single additional rule with one new nonterminal.

For recognition of the defined line structures proved to be robust: the presence of continuous lines and letters did not disturb the results. Applications to other types of documents, like road maps, confirm this result. The Prolog implementation allows rapid evaluation and adaptation

96 MODEL GUIDED INTERPRETATION [Ch. 7

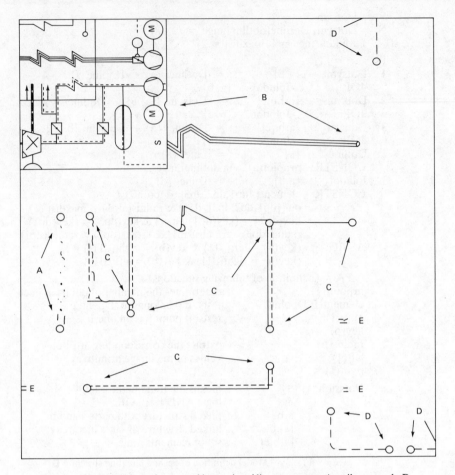

Fig. 4 — Original binary image and instantiated line structures: A: railway-track, B: three parallel lines, C: dashed, E: short parallel elements.

of rules and parameters. The overall runtime of the recognition is about a factor of 5 slower than our implementation in Pascal. But the time to implement the rules is considerably shorter and compensates the slower runtime.

7. REFERENCES

[1] Bartenstein, O., Maderlechner, G., Die Methode der diskriminierenden Graphen zur fehlertoleranten Mustererkennung, Mustererkennung 1984, DAGM/OeAGM Symposium, Springer Verlag 1984, 222–228.
[2] Kuner, P., Efficient Techniques to solve the Subgraph Isomorphism Problem for Pattern Recognition in Line Images, Proc. 4th Scand. Conf. on Image Analysis, Trondheim, June 1985, pp. 333–340.
[3] Klein, F., Kübler, O., Applications of Distance Transforms — Model Guided Interpretation, submitted to *Pattern Recognition*.

[4] Tang, G. Y., Huang, T. S., "A Syntactic-Semantic Approach to Image Understanding and Creation", *IEEE-PAMI*, Vol. 1 (1979) 135–144.
[5] Davis, L. S., Henderson, T. C., "Hierarchical Constraint Processes for Shape Analysis", *IEEE-PAMI*, Vol. 3 (1981) 265–277.

8

Modula– –Prolog — A programming environment for building knowledge-A systems

Carlo Muller, Brown Boveri Research Center, Artificial Intelligence Group, CH-5405 Baden, Switzerland

1. INTRODUCTION

Software engineers in industry and research find themselves confronted with an increasing number of problems that can only be solved with the aid of specialized knowledge and advanced reasoning methods. This means that knowledge must be formalized, stored and made accessible to a deductive component. Applications of this kind are known as knowledge-based systems. These are for instance: diagnosis of technical components, planning and control of processes in various domains, configuration of complex systems, analysis and intepretation of large amounts of data, etc.

Many sub-problems encountered in these areas can or must be solved with the conventional procedural methods. However, the representation and manipulation of the underyling knowledge requires techniques from the domain of applied artificial intelligence.

Modula-2 [1] is a general-purpose procedural programming language, suitable for both systems and applications programming. It is based on the module concept and offers powerful constructs for specifying procedural knowledge (algorithms). Prolog [2] is a declarative computer language, used for solving problems that can be expressed by static symbolic knowledge, i.e. by facts and rules describing objects and their relationships.

We do not attempt to prove the superiority of the procedural over the declarative approach or vice versa. The goal of the following sections is to show that Prolog and Modula-2 have complementary properties which together constitute the basis for a powerful software development tool.

Modula– –Prolog [3] is a language system which integrates the properties of both Prolog and Modula-2. It is not a new programming language with its own syntax and semantics; on the contrary, it supports separate programming in Prolog and Modula-2 and is capable of smoothly interfacing program parts written in the two languages. Modula– –Prolog was designed as an implementation tool for applications requiring procedural knowledge as well as deductive reasoning on static knowledge.

Modula– –Prolog is adequate to perform the first steps in the field of knowledge-based systems. The Prolog component encourages experiments with symbolic programming and AI methods in general, whereas the Modula-2 component supports the conventional procedural programming methods.

Section 2 of this paper shows that Modula-2 and Prolog have complementary properties. Section 3 describes the suitability of Modula– –Prolog for knowledge-based problem-solving. A detailed description of Modula– –Prolog is given in section 4. A short overview of projects that are based on Modula– –Prolog can be found in section 5.

2. PROLOG AND MODULA-2: A SURVEY OF PROPERTIES

Programming style

A computer program can be considered as a coded representation of knowledge. In conventional applications this knowledge is mostly procedural, i.e. it consists of sequences of operations to be executed by the computer system. A typical Modula-2 program explicitly specifies what actions must be performed in what situation.

The declarative programming language Prolog supports a completely different programming style: the programmer declares facts and rules about the objects involved in a problem. This static knowledge is then used to solve the problem without the need for a detailed list of instructions (Fig. 1).

The procedural and declarative approaches to programming are very different, but not exclusive. Most procedural languages support a simple form of declarative programming: an arithmetic or boolean expression is a small declarative program that does not specify how the result is to be computed. Indeed, the scheme of evaluation may differ among different languages (e.g. boolean expressions in Pascal and Modula-2 are evaluated differently). The fact that a declarative program has to be executed in one way or the other imposes a procedural meaning on such a program. The knowledge about the procedural semantics of Prolog is absolutely necessary for writing correct and efficient programs.

Execution speed

Running a declarative program is very costly: a Prolog system spends most of its time searching for solutions that are consistent with the program, i.e. with the facts and rules currently in the Prolog database. In a real-life application, the search space can become very large. In this case the ease of programming in Prolog must be paid by a slow program execution.

The execution speed of Modula-2 programs, on the other hand, is very high. Procedural languages are efficiently translated into the machine

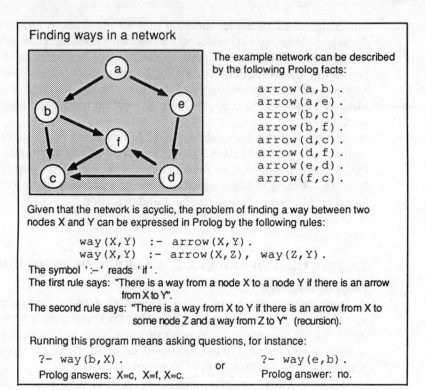

Fig. 1 — A simple Prolog programm.

language of the host computer. The drawback of procedural programming is the tedious task of finding an algorithm solving the given problem, and to specify the correct control structures necessary to execute the algorithm.

Rapid prototyping

The first step in the problem solving process is very often the formulation of a model that describes the problem domain and the involved objects. At this early stage of the program design, the model is only present in a vague form in the programmer's mind. The rapid realization of a prototype demands a flexible tool that supports the formalization of the model and allows an incremental design with executable intermediate results. Prolog is a powerful tool in the domain of rapid prototyping: it is highly interactive and suppports incremental problem solving. At each stage of the model-building process the programmer has an executable specification of the model, i.e. he can evaluate each design decision immediately.

In Modula-2, a different programming style is required. In order to write readable and easily maintainable programs, the specifications must be completed and the algorithms must be understood before programming starts. The program development cycle consisting of editing, compiling, linking and testing normally takes a long time. A procedural, compiled language is not suited for rapid prototyping because slight changes in the specifications often require a complete redesign of a program.

Knowledge representation

The number of applications requiring the formalization of large amounts of knowledge is rapidly growing. The Prolog syntax is very flexible and lends itself particularly well to knowledge representation. Lists and arbitrarily nested dynamic structures are included in the language. New prefix, infix or postfix operators can be declared in order to extend the syntax recognized by the parser (Fig. 2).

It is very easy to design a problem-specific knowledge representation language in Prolog, without the need for writing a separate parser (Fig. 3).

Modula-2 does not provide standard dynamic data structures such as lists. Instead, it offers tools for constructing them out of basic primitives (i.e. standard data types, records and pointers). The representation of dynamic knowledge in Modula-2 is tedious and cumbersome: due to strong data typing, the definition of a record cannot be changed during run-time of a program, but the program-source must be updated, re-compiled and re-linked. Experiments with data structures are therefore very time-consuming.

Program libraries

Many standard problems from all the domains where computers are in use have already been solved in a procedural language and can easily be translated to Modula-2. Large program libraries and collections of efficient algorithms are known today and are available on most computer systems. This is not true for Prolog. The language exists for more than ten years, but still the number of widely available applications is very low. Many of these applications are either experimental or not available on a large number of computer systems.

Communication with the external world

The facts and rules that make up a Prolog program, describe a small world of objects and their relationships. Running the program means asking questions about this restricted world. A Prolog program is thus a closed system, unable to communicate with the external world, i.e. with the operating system of the underlying host computer. The exceptions are some

> An operator is defined by three properties:
>
> - the position: prefix, infix, postfix.
>
> - the precedence class, represented by an integer:
>   operators with a low precedence number bind their arguments stronger than those with higher numbers.
>
> - the associativity: left, right.
>   the associativity disambiguates expressions containing two operators with the same precedence.
>
> Examples
>
> | operator | position | precedence | associativity |
> |----------|----------|------------|---------------|
> | + | infix | 31 | left |
> | * | infix | 21 | left |
> | # | infix | 50 | right |
>
> | Operator notation | Standard notation | Tree representation |
> |-------------------|-------------------|---------------------|
> | a + b * c + d | +(+(a,*(b,c)),d) | |
> | a * b * c | *(*(a,b),c) | |
> | a # b # c | #(a,#(b,c)) | |

Fig. 2 — Operators in Prolog.

predefined built-in predicates. Typically these predicates perform input and output operations as side effects, but in general they are very low level. For example, the predefined predicate **tell** can be used to change the current output stream to a file instead of the computer terminal's screen. Subsequent output generated by other built-in predicates is then added at the end of the specified file.

Today, many computer systems offer advanced input and output capabilities, such as windows, bit-mapped graphics, mouse input, etc. Normally, Prolog cannot take advantage of these features.

This is different in the procedural world. In most implementations of procedural languages, operating system routines can be called and low level

Representing frames in Prolog

The following operators must be declared:

operator	position	precedence	associativity
:	infix	50	right
.	infix	51	right
--	infix	52	right

Example of a frame (taken from a diagnosic system for analog print boards):

```
physical_object
    -- kind
        . value : class
    -- mount_position
        . member_slot : yes
        . default_value : [0,0]
        . comment : '[x,y] position on a print board'
    -- physical_view
        . member_slot : yes
        . if_needed : draw_symbol
```

Representing a production rule in Prolog

The following operators must be declared:

operator	position	precedence	associativity
of	infix	30	right
=	infix	40	right
cf	infix	130	right
and	infix	145	right
then	infix	149	right
if	prefix	150	-
:	infix	175	right

Example of a production rule (cf means certainty factor):

```
rule_4:
    if  has_sauce = yes
        and sauce of meal = spicy
    then body of wine = full cf 0.9
        and special_characteristics of meal=spiciness cf 1
```

Fig. 3 — Knowledge representation in Prolog.

tasks can be programmed directly. The access of external devices and procedures is thus guaranteed. In a large number of Modula-2 implementations, these features are embedded in separate system-modules.

Some properties are complementary

The above survey points out that Prolog and Modula-2 have complementary properties. Many of the drawbacks of one language are strong points of the other (Fig. 4). Modula--Prolog combines both languages in one system so that a programmer has an adequate tool for each part of the problem to solve.

	Prolog	Modula-2
Execution speed	−	+
Knowledge representation	+	−
Description of algorithms	−	+
Rapid prototyping	+	−
Program libraries	−	+
Deductive reasoning	+	−
Interfaces to the external world	−	+

Fig. 4 — Complementary properties of Prolog and Modula-2.

3. KNOWLEDGE-BASED PROBLEM SOLVING WITH MODULA– –PROLOG

Commercially available products for knowledge-based problem solving range from interpreters and compilers for symbolic programming languages up to specialized expert system shells. Very often, these tools do not support system programming tasks and procedural knowledge representation (i.e. algorithms) sufficiently.

Most expert system shells require experience in the field of knowledge engineering. In order to exploit such a shell in practice, the techniques of knowledge representation must be deeply understood. The KEE™ system for instance (KEE is a trademark of IntelliCorp.) expects the familiarity with concepts such as *frame, inheritance, object-oriented programming, access-oriented programming* or *production rule* [4]. Every user of such a powerful but complex system needs extensive training before he can develop an application of a reasonable size.

High costs of a commercial expert system shell are only justified if the structure of the problem at hand is already known and compatible with the type of knowledge representation provided by the shell. Programmers who are not familiar with knowledge-based problem solving usually do not have enough experience to decide what type of knowledge representation and what inference mechanism is best suited to their problem.

In an industrial environment it might be difficult or even impossible to perform the step from the conventional hardware environments and the procedural methods of programming, to the symbolic oriented expert system shells and the special hardware which is sometimes required (for instance Lisp machines). The industry meeds a tool that allows a smooth transition from the conventional to the knowledge-oriented AI methods, and a means of combining both domains.

Modula– –Prolog is adequate to perform the first steps in the field of knowledge-based systems. The Prolog component supports symbolic programming, deductive reasoning, knowledge representation and AI methods in general. Due to the Modula-2 component, all the experiences with the conventional programming methods are still of value, and the procedural aspects of industrial applications are fully supported. The possibility of writing programs partly in Modula-2 and partly in Prolog increases the flexibility and versatility of a prototype knowledge-based diagnostic system that could be realized with Modula–Prolog: the Prolog component is responsible for the representation and manipulation of the underlying knowledge; the interface to existing hardware components and data bases is realized via Modula-2.

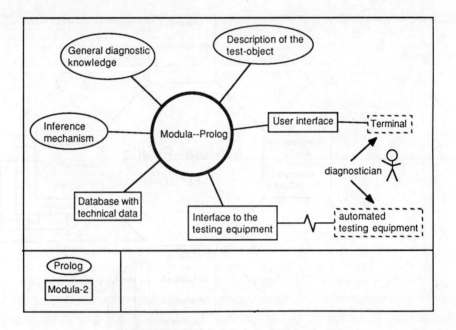

Fig. 5 — Knowledge-based diagnosis of technical components, based on Modula–Prolog.

4. MODULA– –PROLOG IN DETAIL

Modula– –Prolog is a software package written in Modula-2. It consists of a highly portable Prolog kernel which interfaces in many ways to other Modula-2 programs. The interface is designed as a library module which provides all the functions necessary to call the Prolog interpreter from another Modula-2 program, or to include new built-in predicates in Prolog, written themselves in Modula-2. Together with a small user interface, the kernel represents a complete Prolog interpreter, fully compatible to the Prolog standard described in the textbook by Clocksin and Mellish [2].

The Prolog kernel is highly portable because it requires only a minimum of functionality provided by the underlying hardware and operating system. It uses simple sequential I/O operations (streams) to communicate with files, and terminals, and can therefore be implemented on virtually any computer and operating system with a Modula-2 compiler.

The purpose of the interface to Modula-2 is to allow a modular expansion of the Prolog kernel. It offers great flexibility in the construction of individual building blocks that fit together to form a highly specialized Prolog environment, exploiting all the capabilities of the given hardware and supporting application-specific system programming tasks.

Fig. 6 illustrates the expansibility of the Modula--Prolog kernel. The

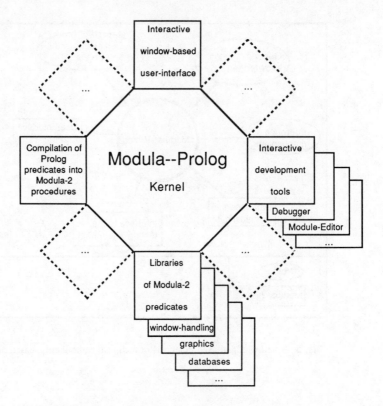

Fig. 6 — The expansibility of the Modula–Prolog kernel.

square boxes surrounding the kernel designate building blocks that are currently in use or under development at the Brown Boveri Research Center on a VAX/VMS computer system.

In Modula--Prolog the basic Prolog functions, such as the parser, or the prover are isolated and can be called separately as library procedures from various Modula-2 programs. In order to execute a Prolog query, the calling program passes a string to the parser which returns an internal representation of the corresponding Prolog term. This term is then used as input for the prover. If a solution exists, the calling program can access the variables of the query, now bound to Prolog terms containing the results of the proof. The Modula-2 program can read these terms and use their values in further computations or just print them out on the screen.

The calling program need not know the internal representation of terms. A Prolog term is therefore declared as a hidden type in Modula-2. Modula--Prolog provides two sets of procedures for processing terms: term-assembling procedures and term-disassembling procedures. These allow both the construction of a term out of its basic components (atoms, functors, numbers, variables) and the separation of the basic components of a Prolog expression for further use in Modula-2. A term in Modula- -Prolog can be any legal Prolog expression including lists and arbitrarily nested structures.

Input and output can be redirected to user-specific procedures so that for instance windowing functions on the host computer system can be exploited. It is possible to redirect trace output, standard Prolog output and output generated by the listing components. Modula- -Prolog may be adapted to a variety of screen and window packages. Fig. 7 shows a hardcopy of a typical

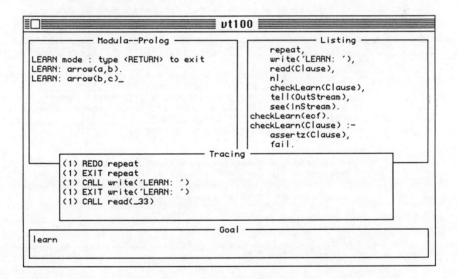

Fig. 7 — Hardcopy of a Modula–Prolog screen.

Modula- -Prolog screen generated with a VT100 terminal emulation program on a Macintosh connected to a VAX.

Modula- -Prolog also provides features for extending the set of built-in predicates, i.e. it is an instrument for building powerful user-tailored Prolog systems which communicate with Modula-2 procedures. Integration of graphics or database operations in the Prolog language are such possible extensions. The user-defined built-in predicates can manipulate Prolog terms in their full generality by using the term-assembling and termdissassembling procedures. The writer of a built-in predicate can call the Prolog unification procedure to perform variable bindings and consistency checks. He has full control over the backtracking behaviour of his predicate, i.e. the procedure can produce multiple solutions when backtracking occurs. For Prolog programmers, there is no difference between a standard built-in predicated or a user-defined Modula-2 predicate.

Fig. 8 lists the Modula-2 source of a user-defined built-in predicate. The

```
PROCEDURE member (VAR u: UserRecord);
VAR list, element, listElement: PrologTerm; (* hidden type *)
    done: BOOLEAN;
BEGIN
  (* isolate the second argument from the goal term 'u.term',
     and store it in 'list' *)
  GetArgument (u.term, 2, list);
  IF (u.count = 1) THEN   (* it is the first call *)
    (* check if 'list' is really instantiated to a Prolog list. *)
    IF (TypeOfTerm (list) <> List) THEN  (* it's not a list *)
      u.proved := FALSE; (* goal fails *)
      u.redo := FALSE; (* no backtracking *)
    ELSE
      (* initialize a global index for backtracking. *)
      globalIndex := 1;
    END;
  END;
  (* search the list for elements that match the first argument *)
  LOOP
    (* isolate the element that is at position globalIndex in the
       list and store it in 'listElement' *)
    GetListElement (list, globalIndex, listElement, done);
    IF NOT done THEN  (* globalIndex too large: end of list reached *)
      u.proved := FALSE; (* goal fails *)
      u.redo := FALSE; (* no backtracking *)
      EXIT;
    ELSE
      (* isolate the first argument and store it in 'element' *)
      GetArgument (u.term, 1, element);
      IF Unify (listElement, element) THEN (* the elements match *)
        u.proved := TRUE; (* the goal succeeds *)
        u.redo := TRUE; (* backtracking enabled *)
        INC (globalIndex);
        EXIT;
      ELSE
        INC (globalIndex);
      END;
    END;
  END; (* loop *)
END member;
```

Fig. 8 — An example of a Modula-2 predicate.

code is not documented in detail, but it gives some insight into the mode of operation of the Modula- -Prolog interface. The predicate is called **member** and has 2 arguments. The second argument must be instantiated to a Prolog list. The predicate succeeds if the first argument is an element of this list. The definition of the equivalent predicate in pure Prolog looks as follows:

member (X, [X|]) .
member (X, [|Y]) :- member (X,Y) .

The predicate can be called in the following ways:

?- member (c, [a,b,c,d]) .
Prolog answer: yes.
?- member (X, [a,b,c,d]) .
Prolog answer: X=a; X=b; X=c; X=d.

The execution speed of the Modula-2 predicate **member** is of course much faster than that of the equivalent Prolog predicte.

5. EXPERIENCE WITH MODULA--PROLOG

The Modula--Prolog package is used in many projects in industry and research. It is part of the "Smart Data Interaction package" [5], an interactive software tool for deductive data manipulation. This package allows efficient access to data on secondary storage, deductive queries and data manipulations with Prolog, and effective human–computer interaction. It is based on the Grid File [6] data structure to provide multi-key access, the Easy human–computer interface and Modula--Prolog which plays the role of an "intelligent" mediator between the user (i.e. Easy) and the data (i.e. the Grid File).

Modula--Prolog is also the basis of several knowledge-based systems that are currently under development. The domains of application are: 3-dimensional geometric construction, search for optimal paths, diagnosis of analog circuit boards, scheduling and planning, etc.

Portability was a major design goal of Modula--Prolog, therefore it is based on a portable operating system interface and utility library for Modula-2, called OSSI. OSSI defines a virtual machine that can be implemented on any computer system where a Modula-2 environment is available. Modula--Prolog has been installed on VAX under VMS, on Unix systems with Modula-2 compiler, on personal computers under MS-DOS and on many systems based on the Motorola 68000. Detailed information on the package can be obtained from the author.

ACKNOWLEDGEMENT

The author developed the Modula--Prolog package at the Swiss Federal Institute of Technology (ETH) in Zürich, with partial support of the Brown Boveri Research Centre. I would like to thank Prof. J. Nievergelt who initiated and supervised the project at ETH. In am also grateful to H. Sugaya from Brown Boveri Research Centre for the many fruitful inputs concerning the Prolog interpreter, and to my colleagues E. Biagioni, G. Heiser and K. Hinrichs, co-authors of OSSI and the Smart Data Interaction package.

REFERENCES

[1] N. Wirth, *Programming in Modula-2*, 3rd edition, Springer-Verlag, Berlin, Heidelberg, New York, Tokyo, 1985.
[2] W. F. Clocksin and C. S. Mellish, *Programming in Prolog*, 2nd edition, Springer-Verlag, Berlin, Heidelberg, New York, Tokyo, 1984.
[3] C. Muller, Modula--Prolog User Manual, Research Report KLR 85-107 C, Brown Boveri Research Centre, CH-5405 Baden/Switzerland, August 1985.
[4] R. Fikes and T. Kehler, The role of frame-based representation in reasoning, Communications of the *ACM*, Vol. 28, Nr. 9, pp. 905–920, September 1985.
[5] E. S. Biagioni, K. Hinrichs, C. Muller, J. Nievergelt, Interactive deductive data management — the Smart Data Interaction package, Wissensbasierte Systeme, GI-Kongres 1985, München, Informatik-Fachbericht 112, pp. 208–220, Springer-Verlag, Berlin, Heidelberg, New York, Tokyo, 1985.
[6] K. Hinrichs, Implementation of the Grid File: Design Concepts and Experience, *BIT*, Vol. 25, 1985, pp. 569–592.

9

CONAD: a knowledge-based configuration adviser, built using Nixdorf's Expert System Shell TWAICE

Stuart E. Savory Head of AI R&D Dept., Nixdorf Computer AG, Paderborn, West Germany

1. INTRODUCTORY REMARKS

Artificial Intgelligence (AI) is a phrase liable to (perhaps deliberate or anthropomorphic) misinterpretation as has been pointed out by Bibel (1980). The reader is no doubt aware of stories ranging from the early Golem myth via Mary Shelley's *Frankenstein* and Karel Capek's *R.U.R.* to Asimov's 'laws of robotics' in modern science fiction.

Winston (1979) has stated that the aims of AI are to make computers "smarter" and to research the phenomena of natural intelligence; to this end AI is defined here as being a (still heterogeneous) collection of computer supported techniques emulating some of the natural capabilities of human beings. This paper takes a much more pragmatic view, and describes some of the current work emerging from industrial AI laboratories (as exemplified by the author's laboratory), which is intended to solve problems of real commercial interest having a significant economic payback.

2. SOME MEDIUM-TERM AIMS OF INDUSTRIAL RESEARCH IN ARTIFICIAL INTELLIGENCE

As stated above, AI is a (still somewhat heterogeneous) collection of techniques (principally search and inference) used in a variety of research areas, including:

1. Expert systems
2. Knowledge based learning and teaching systems
3. Natural language interface
4. Speech output and input devices
5. Robotics and computer vision

6. Program provers
7. Fully automatic natural language translation
8. Completely autonomous robots for hostile environments.

This sequence represents my personal opinion with regard to the sequence in which commercially viable products will become available in the marketplace. The first four points are the subject of our intermediate-term research and development.

We believe that by the next decade we will have made significant progress towards computer systems with which (any) users can converse in (continuously spoken) natural language accessing high-level expertise to solve their particular problem. Experts will be able to teach their expertise to an expert system shell via knowledge acquisition interfaces which construct situational rules by induction from examples (so-called machine learning). Such shells will subsequently be able to use this knowledge for problem solving, and when operating in a pedagogic mode impart the knowledge to (human) students, modelling the students' particular educational levels and idiosyncrasies. The same knowledge bases will be accessible in different languages merely by changing the lexica, grammars and parsers as required.

Today's reality is a first step towards this vision. My team has produced a generic expert system building tool called TWAICE which has been on the market since the end of 1984. It has been used to build a number of expert systems; to keep the length of this chapter down I have outlined one of our internal applications (CONAD) only. We are also building a set of tools for improved knowledge acquisition, some of which were shown at the Hannover trade fair. These tools include consistency and completeness checkers as well as induction tools. Other projects include natural language interfaces accessing relational databases and expert systems (these interfaces have to cope with ellipses, anaphora, indirect speech acts etc.); generating speech output from text in several different languages; recognising spoken input; and the construction of a parallel Prolog hardware.

3. TWAICE: A TOOL FOR BUILDING EXPERT SYSTEMS

Someone is considered an expert if he has a large knowledge domain available in the form of facts and the inferencing rules concerning appropriate application of those facts, and, in addition, has individual experience generally not found in the literature of the domain. Such experience consists of heuristics, analogies, judgements made on the basis of individual decision criteria etc. This experience and knowledge enables the human expert to choose promising problem solving strategies or — if these turn out not to be successful — to backtrack to the point where the strategy failed and try another alternative.

Expert systems are software and databases which store expert's factual and inferential knowledge. They use heuristics and vague knowledge and are able to draw conclusions from given data. In addition, they have an explanation component which can inform the user at any point of the solution which hypotheses they are pursuing, why they chose a particular strategy, what conclusions were drawn so far, and why these conclusions were drawn. The necessity of having a good explanation capability was stated as early as two thousand years ago (Virgil ca. 40 BC), I quote, "Felix qui potuit rerum cognoscere causas", which, if my schoolboy Latin still stands me in good stead, means "Happy is he, who has availed to know the causes of things".

Expert systems shells must be generic, that means independent of the particular knowledge domain used, if they are to make economic sense (i.e. reusability). TWAICE has achieved this aim. It has the following outstanding features:

— An explicit taxonomic model for the knowledge domain(s), in order to have the advantages of "strong typing" not provided by Prolog or Lisp.
— Interactive incremental acquisition of the taxonomic structures from the expert or knowledge engineer
— Interactive incremental acquisition of production rule knowledge from the expert or knowledge engineer
— A production rule compiler for added speed at runtime and for taxonomic consistency checks at compile time
— Multiple possible inference mechanisms including:

  — Backward chaining production rules
  — Forward chaining production rules
  — Confidence ratings for imprecise knowledge
  — Frame represenatation of taxonomic constraints
  — Intra-frame inheritance for knowledge transfer
  — etc.

— A dialog runtime interface which is easy to use, having *inter alia* the following features:

  — Automatic generation of any questions to be posed to the user
  — Choice of dialog language (e.g. subsets of German or English etc.)
  — An "ask-first" capability for gathering initial data
  — A "Help" capability to show the expected (range of) answers to any question
  — Automatic spelling correction and expansion of abbreviations
  — Knowledge-based input validity checks
  — Taxonomy (subset) inspection capability
  — Rule (subset) inspection capability
  — Ability to cope should the user's reply be "don't know"
  — Inspection of the facts deduced so far (or a subframe thereof)
  — Rule based formatting for output of results
  — Optional expert-designed paraphrases of tricky questions

— Offloading of human short-term memory, by restating the current context after an aside

— Explanation capabilities including the ability to ask the expert system at any time during runtime any of the following:

  — Why it asks a specific question
  — How it deduced a given fact
  — Why another conclusion was not reached
  — How a specific conclusion could possibly be reached
  — Which rules are relevant in a given context or to a given frame
  — What the taxonomic relationship of a given frame or slot are
  — etc.

— Facilities for including the user's "attached procedures" stated directly in Prolog or C etc.
— Suport capability for thousands of rules, multiple data tables etc.
— Connection to the user's usual text editor for text processing
— Logging of case histories of individual or multiple sessions for later (knowledge base debugging) analysis by a knowledge engineer
— Full or selective inference tracing, to alow the knowledge engineer to observe and debug the knowledge base at runtime
— Test case management as a debugging aid
— Mathematical formula manipulation
— A library of attached procedures for, e.g., statistics
— etc.

The list above (by no means exhaustive) is intended to give an outline of the features that one can expect of a modern expert system building tool.

TWAICE was two years in development. The current release is version 2.4; it is written in a portable subset of Prolog (see Noelke 1984), and runs on the Nixdorf 8890 and IBM 370 (under VM) as well as the Nixdorf TARGON/32 fault-tolerant family, the TARGON/35 supermini, the M68020 based TARGON/31 Supermicro (all under UNIX), the Tektronix 4404 and DEC VAXes under UNIX and/or VMS. It comprises over 8000 lines of source code, and occupies some 455 kilobytes of memory. To this must be added 261 kB for the Prolog interpreter and so 716 kB is neded for the basic shell. The compiled version is smaller and of course faster. On top of this comes the knowledge base; a medium-sized (400 rule) knowledge base required some 338 kB, working memory for the deduced facts (30 kB) and some 800 kB for runtime stacks and the heap. Thus a typical consultation with TWAICE requires 2 megabytes of (virtual) memory.

On the Nixdorf TARGON/35, which is a 3 MIPS machine, the average response times are 0.8 seconds for the user, and ca. 2 seconds when in knowledge engineering mode, both with a working set of just on 1 megabyte. On a VAX 11/780 you would have to triple these response times.

Implementation details about TWAICE and the knowledge acquisition tools being built to support machine learning can be found in Savory (1985). The basic paradigms of rule-based expert systems are given in Buchanan *et al.* (1984).

4. CONAD: A KNOWLEDGE-BASED CONFIGURATION ADVISER

CONAD is an expert system built using the TWAICE shell. It comprises over a thousand rules which are used to configure the Nixdorf 8864 family of banking computers. Due to the modularity of this family there is a combinatorial explosion of configuration alternatives, not all of which are feasible or realistic. In the past, sales personnel would sometimes generate orders which were self-contradictory, incomplete, contained superfluous parts, did not fulfill performance requirements, were non-price-optimal or were otherwise inadequate, CONAD is used to acquire order data in dialog, generating orders which are consistent, complete, not oversold, meet performance requirements, correctly priced, buildable and installable in the desired location. It thus fullfils the same purpose as R1 (McDermott 1984), but does so in a different manner. A first toy prototype which preceded TWAICE was built and demonstrated in 1982/83 and is described in (Savory 1983).

In a typical CONAD session some 74 questions are asked of the user (some might be answered by a direct access to the customer configuration logfile) from which on average 517 facts are deduced, whilst configuring banking system with nine different terminal types, printers, discs, telecommunications equipment, ID-card readers, passbook printers, automated tellers, line encryption etc. etc.

The following (constructed) example demonstrates the flavour of the system showing its deductions:

(Q 4) In which country is the system to be installed ?
-->> GB.
Rule 273 deduces mains voltage is 240 volts.
Rule 273 deduces power frequency is 50 hertz.
Rule 422 deduces power fuse is rated at 13 amps.
Rule 422 deduces power plug type is UK norm.
Rule 111 deduces QWERTY keyborads are default.
Rule 977 deduces there are 2 decimal
                positions in currency fields.
Rule 355 deduces Lloyd's security measures apply.
   etc.

In realworld applications, as opposed to the example above, the questions asked of the user tend to concern themselves with details of variants of customers end-user devices, whilst the deductive mechanisms make the major decisions. Since only necessary questions are asked of the CONAD user, it is not possible to configure impossible machine setups, which is one of the disadvantages of serial batch runs as used, e.g., in the XSEL/R1 system.

CONAD now has over 1500 rules and took the expert (Werner Koehler) some fourteen months elapsed time to develop. Current productivity with TWAICE far exceeds this figure, since CONAD was developed in parallel with TWAICE, which meant that the unfortunate expert did not have a stable base to work with, as we do now. For example Professor Krallmann (TU Berlin) and his students have Built a 700+ rule system in just 3 months (cf. Mensel & Michel 1985).

REFERENCES

Bibel, W. (1980), Intellektik statt KI, *Rundbrief der Fachgruppe KI in der GI*, Vol. 22, 1980, pp. 15–16.

Buchanan, B. and Shortcliffe, E. (Eds.) (1984), *Rule based expert systems*, Addison-Wesley.

McDermott, J. (1984), R1 revisited: four years in the trenches, *The AI Magazine*, Fall 1984, pp. 21–32.

Mensel, G. and Michel, J. (1985), Development and construction of a knowledge-based consultancy and configuration system in the sector of stock- and production-control, Proceeding of the COMPAS '88, VDE Verlag.

Noelke, U. and Savory, S. E. (1984), PROLOG-Systeme im Vergleich, *Applied Informatics*, 3/84, pp. 108–112.

Savory, S. E. (1983), The prototype Nixdorf Expert System, *Applied Informatics,* 11/83, pp. 478–482.

Savory, S. E. (Ed.) (1985), *Kuenstliche Intelligenz und Expertensysteme*, Oldenbourg Verlag (Munich).

Virgil (70–19 B.C.), "Georgics", ii, 490.

Winston, P. (1979), Lecture given during the international AI summer school, Dubrovnik, August 1979.

10

A Prolog Frame System for Knowledge-based Design and Diagnosis

H. Sugaya, Brown Boveri Research Center, Artificial Intelligence Group, CH 5405 Baden, Switzerland

1. INTRODUCTION

The design and diagnosis of technical systems have been treated often as separate activities. The design activities involve synthesis of a system satisfying a set of various conditions: functionality, performance, extensibility, testability, price, etc. The final system must be described through functional modules and their interconnections. The diagnostic activities involve analysis of a system investigating the causal relationship between faults and symptoms based on data about the system structure and functionality as well as on some heuritics. This paper reports on an ongoing research in which a prototype of a knowledge-based program is being implemented for design, simulation, and diagnosis of technical systems.

Traditionally, computer aided design applies simple but general algorithms to solving design problems under a set of well-defined constraints. Major design decisions are made by the designers prior to the computation by programs. One possible aid to this aspect of design activities is a program which possesses a large collection of specific knowledge that can be used for decomposing a goal function into subsystems. Central to this activity is the modelling of a system through structural and functional descriptions in consideration of design rules and constraints. The structural and functional description permits simulation of the modelled system to evaluate the soundness of design decisions.

Diagnosis has been considered as an ill-structured problem in medical field. Traditionally, heuristics have been applied to the modelling of such structures. However, technical systems when appropriately designed (we mean designed for testability) can be systematically diagnosed using their structural and functional descriptions [Davis *et al.* 82], [Genesereth 82]. Heuristics can be applied only on demand. This contrasts with the graph-theoretic approach for diagnosis (see for example [Duhamen and Rault 79] for the treatment of automatic test generation techniques). As with the design, diagnostic activity requires structural and functional descriptions of a system. Simulation is necessary for computing correct input/output values to be compared against measured values. Design rules stored in the knowledge base can also be used for explanation.

Our thesis is that a program which accepts structural and functional descriptions of a technical system can aid the design and diagnostic activities when these descriptions are augmented with domain specific knowledge such as design rules and constraints. To test out our thesis, we have chosen a particular domain, namely the design, simulation and diagnosis of analog circuits since a knowledge-based approach requires a broad store of specific data. Another criterion was to select an application area in which the design and diagnostic activities involve some deep reasoning. Analog circuits contain hybrid components with electronic, electrical and mechanical properties. Industrial automation systems, process control systems and turbine generators are but more complex examples of such hybrid systems.

One of our research goals is to construct a knowledge programming environment in which a designer decomposes an analog system into subsystems interactively satisfying both constraints and functionality. This process is repeated until a subsystem is found, which is already designed, or a component is reached, which is a smallest unit of design or repair. Design rules are applied wherever possible, but major decisions may still be made by the designer. Another goal is to investigate the representation of structural and functional knowledge in the frame language (rules for design and diagnosis might be part of the functional knowledge). The experimental system contains two hierarchies of concepts for structural and functional descriptions. These descriptions contain both an abstract schema for a prototype of its instances. The descriptions also contain a set of demons for resolving inconsistencies and generating a member of the class according to the design rules being applied.

The prototype system uses a data base of knowledge to store structural and functional descriptions as well as rules for design and diagnosis. The data base is used to provide factual data, missing details, maintain consistencies and compute requested values. These declarative and procedural knowledge is represented in a Prolog frame language — an implementation of the frame concept [Minsky 75] in Prolog similar to FRL [Goldstein 77] and Units [Stefik 79].

Frame-based knowledge programming systems in Lisp have evolved over the past ten years, for example Loops [Bobrow and Stefik 81] and KEE [Fikes and Kehler 85]. Being influenced by the Smalltalk system [Goldberg and Robson 83], they integrate several programming paradigms: object-oriented, access-oriented, logic-oriented and rule-based [Stefik and Bobrow 85], [Stefik et al. 86]. The same endeavor is also taking place in Prolog, for example Looks [Mizoguchi et al. 84] and Mandala [Furukawa et al. 84]. Prolog has advantages over Lisp-based systems because of the availability of built-in pattern-matching, backtracking and logical variables [Subrahmanyam 85]. ART [Kalme 86] is similar in concepts but differs from our work in that the knowledge is represented by schemas and propositions that are fired by an inference machine. Dedale [Dague *et al.* 85] is most similar to our work in that it incorporates structural and functional descriptions as well as diagnostic rules for analog circuits in frames.

2. MODELLING

In order to design, simulate and diagnose a technical system, it is necessary to have a model of the system. The important elements in modelling is the structural and functional descriptions of the system. Such descriptions readily permit simulation of the designed system. Design task is then to decompose a functional module into submodules satisfying the functionality as well as various constraints such as reliability for correct operations, testability for diagnosis, etc. On-line simulation is a practical tool to test out the soundness of design decisions. Diagnosis task is to investigate the cause-effect relationships through structural and functional dependencies. By functional dependency we mean that each subsystem has its reason to exist for a certain functional role. Again, on-line simulation is a practical tool to compute expected input/output values of a function module to be compared against those that are measured. These three activities being involved in the modelling are illustrated in Fig. 1.

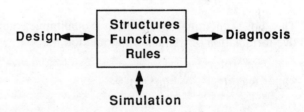

Fig. 1 — Modelling and the three associated activities.

Traditionally, a "circuit schema" is used as an end product of the design work for electronic circuits. Functionality of the circuit is given in a different representation called "function block diagram". The circuit schema contains essentially more detailed descriptions than the function block diagram, e.g. component names, mounting positions, pin connections. However, the designer initially reasons not with the circuit schema, but with the function block diagram. Although both schemas can be used for patent descriptions, the designer's intention can be better understood in the function block schema.

A. Reasoning in logic

In the modelling of an analog component, the structural and functional relationships can be reasoned in logic. We will explain this by using a frequency modulator (Fig. 2). The frequency modulator has two inputs: the

Fig. 2 — Frequency modulator.

signal S_{in} and the carrier frequency f_c, and the modulated output signal S_{out}. The structural dependency leads us to reason that

<S_{out} is correct> IF <S_{in} is correct> AND <f_c is correct> AND <Frequency modulator is correct>.

The correctness of S_{in} or f_c can be formulated in the same way. However, the correctness of the component, i.e. the frequency modulator, requires a different formulation, namely the relation between inputs and outputs. The functional dependency of the component leads us to reason that

<Frequency modulator with f_c, S_{in} and S_{out} is correct>
IF <S_{out} is a frequency modulation of f_c by S_{in}>

The above functional specification is qualitative compared to the quantitative formulation of frequency modulation, i.e.

$$S_{out} = K \cos(f_c + \frac{cM}{\Omega} \sin(\Omega t))$$

where $S_{in} = M \cos(\Omega t)$, and K, M and c are constants. Qualitative formulation expresses an intention of the functional role of the component while the quantitative one defines an input/output relation through numerical equations.

B. Modules, ports and relations

The structural and functional relationship can be modelled by means of *modules and ports*. A module is considered to be a black box with ports for input and output signals. A module may encapsulate components which are again modules. Conversely, modules can be combined together to form a more abstract module. Signals propagate through ports. There ar three kinds of connections between ports: a connection between two modules, a connection into and out of a module depending on the module hierarchy.

Each port can have at most one input and one output connection. Thus, a fan-in and a fan-out are modelled as a module with 1-to-n or n-to-1 ports. Fig. 3 illustrates that the frequency modulator contains three submodules and has two input and one output ports.

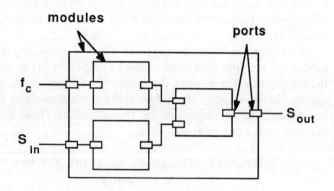

Fig. 3 — Modules and ports.

A module with ports defines a syntactical form of a *relation* between inputs and outputs of a component. We write

$$\text{relation}(Module, I_1, I_2, ..., O_1, O_2, ..., O_n)$$

where I_i are inputs and O_j are outputs. A port binds a signal to or from an argument of the relation. Its semantic is specified as a clause in Prolog so that the specification can be executed for simulation. The relational description has an advantage over functional description because it permits forward and backward reasoning during the design or diagnostic activities.

3. DESIGN, SIMULATION AND DIAGNOSIS

While the concepts of modules, ports and relations are for modelling the structure and function of analog circuits, design rules guide and narrow the possible space of solution. A result of the design is again a model expressed in ports and modules with links to functional descriptions in relations. The prototype system simulates the input/output relation of a module to assert its correctness; if it fails, faults are searched along the connections through ports. Design, simulation and diagnostic activities are elucidated below in connection to the modelling.

A. Design
The most important group of design rules is the decomposition of a goal function into subfunctions. Other design rules are constraints to be satisfied, e.g. for reliable operations, extensibility, testability, economy, etc. These

rules vary in degree for applicability. Rigid rules can be automatically applied for configuration task. Alternative rules may be applied to generate a set of possible decompositions, and one of the rules is chosen according to the constraints. Conflicts may be resolved by giving some priorities to the constraints, or else be done by the designer. Since the number of design rules is infinitely large, we consider only a few of them below. We explain this by refining the frequency modulator for correct operation, flexibility, and testability.

A frequency modulator can be realized by using ring modulator. A ring modulator requires that inputs must be amplified to the impedance level of the modulator for *correct operation*. Such constraint must be satisfied during the decomposition. In Fig. 3, the two submodules at input ports can be interpreted as amplifiers for this purpose. Thus, a rule fo function decomposition can be reasoned as

> decomposition(frequency_modulator, function,
> ring_modulator).

which states the frequency modulator is functionally decomposed to a ring modulator with inputs being amplified. The constraint can be described as

> constraint(ring_modulator,
> correct_operation, amplify_inputs).

which states the ring modulator has a constraint that the input signals be amplified for reasons of correct operation.

The same analog circuits can be designed to cover different carrier frequencies for *flexibility*. One possibility is to use a frequency divider to provide different frequencies. We reason that

> decomposition(frequency_modulation,
> flexibility, frequency_divider).
> decomposition(frequency_modulation,
> function, frequency_modulator).

Note that the two modules frequency modulator and divider are grouped together to form a more general module frequency modulation. In diagnosis, one strategy to localize faults is the "divide and conquer" in which suspective area is roughly divided into half. Since analog signals cannot be measured directly by a probe, an extra circuits is added for reasons of *testability*. We reason that

> decomposition(frequency_modulation, testability, decoupler).

A result of the design satisfying these decomposition rules and constraints is given in Fig. 4. The frequency divider ($\div$ M) provides the carrier frequency f_c. It is then fed to the modulator and also to the decoupler for test output.

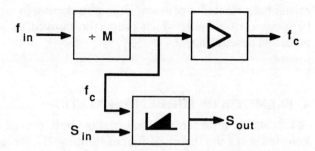

Fig. 4 — Refined frequency modulator.

B. Diagnosis

To derive diagnostic rules, we negate the logical formulation of structural and functional relationships mentioned earlier. We have

> IF <Frequency modulator with f_c, S_{in} and S_{out} is faulty>
> THEN <S_{out} is not a frequency modulation of f_c by S_{in}>.

> IF <S_{out} is faulty>
> THEN <S_{in} is faulty> OR <f_c is faulty> OR <Frequency modulator is faulty>.

The first rule verifies the functional behavior of a frequency modulator. Simulation permits computation of correct values for inputs and outputs. The second rule deduces a fault to components through structural dependencies.

Given a frequency modulation circuit as described in Fig. 4, a fault state of the signal S_{out} can be deduced either by checking the frequency divider or frequency modulator. Assuming S_{in} is correct, testability of f_c permits division of a search area roughly into half. This test can be performed automatically, or a diagnostician may be queried for test probes [Kriz and Sugaya 86]. The latter idea is used to acquire new facts on demand [Sergot 82]. To test the correctness of the frequency divider, we have

> IF < f_c is faulty>
> THEN <f_{in} is faulty> OR <Frequency divider is faulty>.

Causality relationships can be viewed from the point of functional roles [Milne 85]. For example, if S_{out} is fault and f_c is wrong, one may reason that the frequency divider is wrong because it is responsible for providing a correct carrier frequency.

In applying "negation as failures" to logic [Clark 1978], it must be noticed that in diagnosis all elements can be universally quantified, but not existentially treated. For example, some unknown foreign component might be a cause of a fault, which cannot be derived by simply negating the modelled description.

4. ELEMENTS OF THE REPRESENTATION

A knowledge representation must allow both procedural and declarative knowledge. In the theory of frames [Minsky 75], the generalized property lists with annotations are used for representing declarative knowledge, and the procedure attachments for representing procedural knowledge. Knowledge base is organized as hierarchies of concepts in frames. The concepts are frames, and the links are values of slots. The built-in generalization scheme provides a property inheritance to the progeny. Much of the early work on knowledge representation in frames has been done by Goldstein and Roberts [77] in FRL, Bobrow and Winograd [77] in KRL, Stefik [79] in Units, and recently by Fikes and Kehler [85] in KEE, and Stefik and Bobrow [85] in Loops. This section describes the elements of the following five representation techniques being used: class/member frames, inheritance, annotations, composite objects, and perspectives.

Class/member frames

In our modelling application, it is important to separate generic concepts from the concrete descriptions. This distinction permits us to organize knowledge into two kinds: domain knowledge applicable for model description and specific data describing each analog circuits. We use two kinds of frames: member and class. A member frame is an instance of a class frame and contain only the property descriptions of its own. A class frame contains property descriptions of its own and for its member instances. A class frame has two kinds of links superclass and subclass to form hierarchies of concepts. Thus, a member frame points to its class by means of member_of link, and a class frame points to its superclass by means of subclass_of link.

Inheritance

The separation of generic concepts from concrete descriptions allows sharing of knowledge on structuring, functionality of modules and design rules by each instance of circuit components. The class frame permits the hierarchical inheritance of properties in which prototype descriptions are inherited by the progeny. A property description of progeny, by default, overrides the inherited property. Other kinds of inheritance roles are not considered for reasons of simplicity.

Functional knowledge is distributed in a hierarchy of concepts: superclass frames contain properties common to the progeny and subclass frames can refine or add more specific property descriptions. Input/output relation of a function module, for example, can be described as a slot of the class frame, say emitter_follower. An amplifier, which is a member instance of the emitter_follower, may contain a slot for amplitude and inherits the function description of the class frame. The hierarchical inheritance achieves both representational power and storage economy.

Annotations

Annotations to frames are a technique commonly used in the knowledge representation languages. Comments, defaults, type, and procedural attachment (demons) are the annotations to slot values. *Comments* are used do annotate informal knowledge about the slot value, e.g. metric system, hints for design or diagnosis, and debugging aid to a knowledge engineer. The commentary facilitates the system's ability to explain its inferences to a user. *Default* provides either an initial value or a commonly used value. The value facet overrides the default. The utility of defaults mimic "common sense" reasoning out of informal specification. The default utility also facilitates a simulation task. Type facet annotates the form of a slot value. Type can be seen as a generalization of the type concept like in Pascal. Type annotation is used when the value is instantiated or modified. By default, a slot can take a value of any type: it is defined as a Prolog structure. The following types are planned (but not yet implemented): range, one_of, and set_of.

Procedural attachments provide the capability of representing procedural knowledge. Although a procedure can be assigned to a record field in procedural languages, the procedural attachment in frames accompanies a "guard" for when and for what purpose a procedure should be invoked. In our prototype system, procedures are Prolog clauses. Standard guards are combined to the frame access functions, e.g. if_needed, if_be_removed, if_was_removed, etc. Nonstandard ones, and hence be defined by the user, are the input/output relation of a function module as well as design rules for decompositions and constraints.

Composite objects

A composite object is a group of objects that are interconnected together. Composite objects may contain other objects as parts. In our application, a module contains ports as well as modules as parts. For example, the frequency modulator contains a ring modulator with two amplifiers for the impedance matching of f_c and S_{in} (Fig. 3). The recursive definition of modules allows us to model a circuit component of unbounded size and depth.

Composite objects are specified by a class frame. Subparts are instantiated on demand for reasons of flexibility. In our application, ports are created either when a module that contains them is created or when two modules are connected. Similarly, modules that are contained are created when decomposition rules are applied, or else deferred until being specified by a designer.

Perspectives

The composite objects can be interpreted differently depending on a goal to be achieved. A model description for circuits can be viewed as a physical object, function block schema, input/output relation, or causality graph. In our implementation, the perspectives are represented as a specific concept by a class frame and can be inherited by all members. Fig. 6 shows the frequency divider being viewed as a function block which can be dragged to a different place on the screen.

5. IMPLEMENTATION

A prototype of knowledge-based design and diagnosis system has been developed by the author in Modula--Prolog and runs on VAX-11 computers. Modula--Prolog is a Prolog interpreter and supports procedural and declarative style of programming [Muller 85]. The package is also used to realize standard predicates for manipulating window functions. The prototype consists of the following three components: dialog component, frame inference and knowledge base (Fig. 5). The dialog component

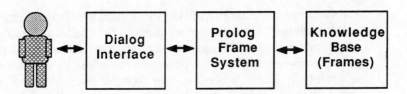

Fig. 5 — System structure.

contains a command interpreter which converts a command into a sequence of calls to the Prolog system. The user interface includes a graphical editor for modelling circuits, frame editor, and simulation kit (Fig. 6.).

Prolog frame system defines a set of access functions for creating, modifying and deleting frames. The access functions can be used with or without demons. The demon function has a form

Demon(InvokingFrame, Value)

Value argument is a bound variable for write access and an unbound variable for read access. In the latter case, the argument is bound to a value after the call. The system also has a set of clauses for (un-) loading and (un-) parsing of frames between the internal memory and an external text file. Frames are internally stored as FrameName(Slot,Facet,Value). Frame-Name is used as a functor only for reasons of efficiency. The internal representation of the frame f_divider is given in Fig. 7. Demon functions

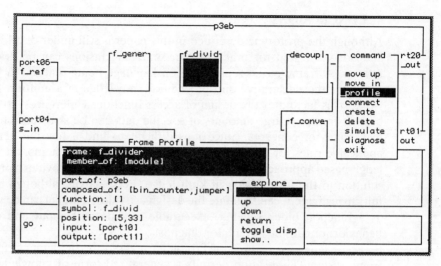

Fig. 6 — The user interface.

```
f_divider( type, value, member).
f_divider( member_of, value, [module]).
f_divider( part_of, value, p3eb).
f_divider( composed_of, value, [bin_counter, jumper]).
f_divider( function, value, nil).
f_divider( symbol, value, f_divid).
f_divider( position, value, [5,33]).
f_divider( input, value, [port10]).
f_divider( output, value, [port11]).
```

Fig. 7 — Internal representation of a frame.

are all Prolog clauses. Input/output relation of the frequency divider, for example, is defined as

r(Module, w(Input), w(Output)):-
    frame_get_vd(Module, divider, K), divide(Input, K, Output).

The predicate w in the argument terms indicates that they deal with input/output frequencies. The predicate divide in the body defines the division as a relation.

6. CONCLUSIONS

Although the prototype described in this paper is still under development and not yet tested for practical use, some conclusions can be drawn as follows. A frame-based representation language being realized in Prolog combines the declarative and procedural knowledge in a uniform framework. This facilitates the design of access functions (inferences) to a data base of frames. Large amounts of specific data can be structured to form hierarchies of concepts. Since it is difficult to find a few but generally applicable rules for the design and diagnosis of analog systems, the knowledge-based approach with a broad store of specific data provides a practical solution to the formulation of knowledge. Perspectives, although not yet fully investigated, can facilitate the design activities: a model can be viewed as a function block schema, executable module with input and output behavior, or causality graph for diagnosis.

Some language features of Prolog have eased the implementation of a frame system. Unification permits a generalized form of knowledge representation to the theory of frames: slots, facets and values are Prolog structures and can be matched against other structures including variables. This facilitates the formulation of annotations in frames (e.g. dynamic arrays can be expressed as slots with index arguments). Backtracking mechanism is used to provide other alternatives. Relational style of programming can be applied to describe input/output behavior of a function module, permitting both forward and backward simulation. Logical variables are used to fix the input/output relation and bound to actual signals during simulation. Furthermore, the direct use of Prolog clauses as an executable specification eliminates a need for designing other logic-based rule formalism with pattern-matching and variables.

REFERENCES

Bobrow, D. G. and T. Winograd. (1977) An overview of KRL, a knowledge representation language, *Cognitive Science,* vol. 1, no. 1, pp. 3–46.

Clark, K. L. (1978) Negation as Failure, in *Logic and Data Bases,* H. Gallaire and J. Minker (Eds.), Plenum Press, New York, pp. 293–322.

Dague P., P. Deves, Z. Zein, and J. P. Adam. (1985) DEDALE: an Expert System in VM/Prolog, ITL meeting on Expert Systems, San Jose.

Davis, R., H. Shrobe, W. Hamscher, K. Wieckert, M. Shirley and S. Polit. (1982) Diagnosis Based on Description of Structure and Function, AAAI-82, pp. 247–410.

Duhamel, P. and J.-C. Rault. (1979) Automatic Test Generation Techniques for Analog Circuits and Systems: A Survey, *IEEE Tr. Circuits and Systems,* vol. CAS-26, no. 7, July, pp. 411-440.

Fikes, R. and T. Kehler (1985) The role of frame-based representation in reasoning, *Comm. of the ACM,* vol. 28, no. 9, pp. 904–920.

Furukawa, K., H. Takeuchi, H. Yasukawa and S. Kunifuji. (1984) Mandala: A logic based knowledge programming, in Proc. Int. Conf. 5-th Generation Comput. Syst., Nov., pp. 613–622.

Genesereth, M. R. (1982) Diagnosis Using Hierarchical Design Models, AAAI-82, pp. 278–283.

Goldberg, A. and D. Robson. (1983) *Smalltalk-80: The language and its implementation,* Addison-Wesley, Reading, Massachusetts.

Goldstein, I. P. and Roberts, R. B. (1977) NUDGE, a knowledge-based scheduling program, 5-th IJCAI, pp. 257–263.

Kalme, C. (1986) AI tool applies the art of building expert systems to troubleshooting boards, *Electronic Design,* April 3, pp. 159–166.

Kriz, J. and H. Sugaya. (1986) Knowledge-Based Testing and Diagnosis of Analog Circuit Boards, The 16-th Int'l Symp. on Fault-tolerant Computing, Vienna, July.

Milne, R. The Theory of Responsibilities, (1985) *ACM SIGART Newsletter,* no. 93, July, pp. 25–29.

Minsky, M. A. (1975) A framework for representing knowledge, in *The Psychology of Computer Vision,* P. Winston (Ed.), McGraw-Hill, New York, pp. 211–277.

Mizoguchi, F., Y. Katayama, and H. Owada. (1984) LOOKS: Knowledge representation system for designing expert system in the framework of logic programming, in Proc. Int. Conf. 5th Generation Comput. Syst., Nov., pp. 606–612.

Muller, C. (1985) Modula– –Prolog user manual, technical report, KLR 85-107C, Brown Boveri Research Center.

Sergot, M. (1982) A Query-the-User Facility for Logic Programming, in *Integrated Interactive Computer Systems,* P. Degano and E. Sandewall (Eds.), North-Holland, pp. 27–41.

Stefik, M. (1979) An examination of a frame-structured representation system, 6-th IJCAI, pp. 845–852.

Stefik, M. and D. G. Bobrow. (1985) Object-Oriented Programming: Themes and Variations, *The AI Magazine,* Winter, 40–62.

Stefik, M., D. G. Bobrow and K. M. Kahn. (1986) Integrating Access-Oriented Programming into a Multiparadigm Environment, IEEE Software, January, pp. 10–18.

Subrahmanyam, P. A. (1985) The Software Engineering of Expert Systems: Is Prolog Appropriate?, *IEEE Tr. on Software Eng.,* Vol. SE-11, no. 11, pp. 1391–1400.

11

Applications of knowledge-based planning systems

Austin Tate, Knowledge Based Planning Group, Artificial Intelligence Applications Institute, University of Edinburgh, 80 South Bridge, Edinburgh EH1 1HN, UK

1. TOPICS TO BE COVERED
— How knowledge-based planners relate to "expert systems".
— Application areas tackled by knowledge-based planners.
— Progress over a 20-year period on knowledge-based planning techniques.
— Currently active research areas.
— An overview of the Edinburgh O-Plan project.

2. INTRODUCTION
General-purpose planning systems which can use application specific information or knowledge to produce plans of action have been a long standing theme of AI research. A large number of techniques have been introduced in progressively more ambitious systems over a long period. There have been several attempts to combine the planning techniques available at a certain time into prototypes able to cope with realistic application domains.

Planning research has often built on earlier work. There have usually been full written accounts of the techniques employed in a form that could be abstracted out from the application domain and re-implemented or incorporated in other systems. A "tradition" of using a simple block stacking domain for examples has aided comparison and improved understanding (though today's more sophisticated planners barely show their true worth in these simple problems). Several researchers have used a set of problems to facilitate efficiency comparisons into the production of realistic prototype planners.

3. EXPERT SYSTEMS AND KNOWLEDGE-BASED PLANNERS

Knowledge-based planning systems share many techniques and problems in common with the rather better known "expert systems".

However, typically the production rule-based expert systems of today are

— used for classification or interpretation
— work in a convergent fashion.

Whereas knowledge-based planning is a "synthetic" task where there is inherently more search since there are alternatives to consider or compare. Planning also introduces its own problems:

— Representing and reasoning about time
— Uncertainty in the execution of plans
— Multiple agents who may cooperate or interfere
— Physical or other constraints on suitable solutions

Many of the knowledge-based planning techniques seek to control the extent of the search. The taxonomy in Fig. 1 shows many of the major planning systems developed in AI and how they relate to one another. A short bibliography is attached to this paper to act as an entry point to the literature. A more comprehensive survey of knowledge-based planning techniques is in Tate (1985).

4. APPLICATIONS TACKLED BY KNOWLEDGE-BASED PLANNERS

Planners have been applied to a variety of application domains (as well as the simple explanatory domains such as block stacking) (see Table 1).

Plans are used as a basic data structure in other AI Applications:

— Story understanding (Schank and Abelson (1977), Wilensky (1978))
— Speech and text generation
— Office systems modelling (Fikes, 1982)
— Intelligent front ends
— Computer-aided instruction
— Program generation

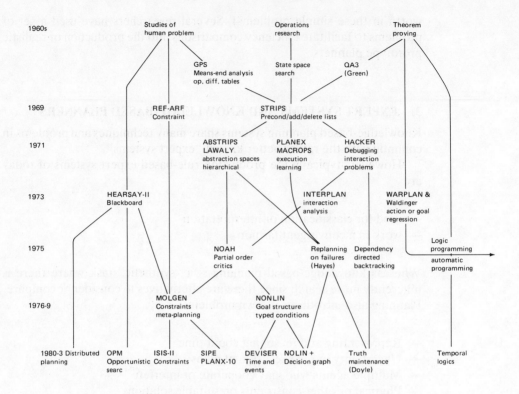

Fig. 1 — A taxonomy of AI planning systems.

Table 1

Domain	Planner
Mechanical engineers apprentice supervision	HOAH
House building and electricity turbine overhaul	NONLIN
Experiment planning in molecular genetics	MOLGEN
Program generation	HACKER, etc.
Electronic circuit design	NASL
Journey planning	OPM (Hayes-Roths)
Job shop scheduling for turbine production	ISIS-II
Aircraft carrier mission planning	SIPE
Naval replenishment at sea	NONLIN
Software production	NONLIN
Voyager spacecraft mission sequencing	DEVISER
Automated factory planning	FORBIN

5. EVOLUTION OF KNOWLEDGE-BASED PLANNING TECHNIQUES

1965 Heuristic search (e.g. Graph Traverser)

State/space heuristic search was an outgrowth of Operations Research branch and bound search techniques.

However, use was made of a state evaluation function that was application specific to guide search to the most promising intermediate states.

1969 World models (e.g. STRIPS)

Symbolic reasoning about states was possible by representing the states in logic (as world models).

It was possible to select relevant operators for the problem by looking for those that changed the world model in appropriate ways.

Operators or actions were modelled by their
   Preconditions — applicability conditions
   add/delete lists — changes to a state world model.

1971 Hierarchical planning (e.g. ABSTRIPS)

Planning at progressively greater levels of detail.

1973 Interacting goals (e.g. regression, goal structure)

Techniques to represent, recognise and deal with conflicting sub-goals.

This led to the evolution of ideas on representing the "intent" of plan steps independently of the detail of the ordering of the actions in the plan.

1975 Partially-ordered actions (e.g. NOAH)

Representing plans as networks of actions rather than strict sequences.

1977 Dependency-directed search

Recording the dependencies between decisions and choices so that failures could be analysed to remove only the invalidated parts of the plan (at plan or execution time).

At the end of this period of evolution, many knowledge-based planners had a similar form that can be summarised as:

(a) based on a hierarchical representation of the plan, a set of goals or some skeleton plan can be given and expanded out to greater levels of detail;

(b) the planner searches through alternative methods of expanding high level plans to lower level ones (and alternative means of

satisfying conditions, choosing objects, etc.); interactions between solutions to different parts of the plan are detected and corrected;
(c) at each level, the plan is represented as a network of nodes in a form that can allow critical path data computations, resource analysis, etc. to prune the search.

6. USE OF CONSTRAINTS TO RESTRICT SEARCH

Later planners have employed this structure and have added facilities to represent progressively more realistic models of the application.

These facilities also augment the application specific information available to the planner to restrict the search space. Hence the use of the term "knowledge-based planning".

Besides improving the range of applications that can be realistically tackled, these facilities have helped improve user explanation, user interaction, etc.

1976 Goal structure (e.g. NONLIN)

Details of why any action is in the plan removes the need to deal with many apparent interactions and search alternatives considered by earlier systems.

1982 Objects as resources (e.g. SIPE)

Ability to recognise conflicting or concurrent use of shared objects.

1982–4 Consumable resources (e.g. DEVISER, NONLIN)

Use of limits on consumables to restrict search and to heuristically weight search alternatives to prefer low consumable use.

1982 Time constraints (e.g. DEVISER, O-Plan)

Ability to specify time windows for gaols or actions, deal with external timed events, etc. Again these constraints can be used to restrict search to solutions that can meet the given time specifications.

1983 Priorities (e.g. ISIS)

Use of preferences and "soft constraints" to guide search.

7. PROBLEM AREAS RECEIVING ATTENTION IN PLANNING SYSTEMS UNDER CONSTRUCTION

— Need to interface within complex, dynamic systems.
— Need to integrate with dynamic condition monitors and ability to be restarted on failure to modify plans.

— Need to use application specific knowledge sources of diverse types.
— Need to cope with much larger problems.
— Need to cope with greater realism in modelling the application.

8. CURRENT RESEARCH TOPICS IN KNOWLEDGE-BASED PLANNING

— Resource usage and constraint handling.
— Execution monitoring and replanning on failure.
— "Opportunistic" and "dependency-directed" search.
— Application knowledge elicitation.
— User interfaces and interactive planning.
— Distributed planning, cooperation and communication or disruption between planners.
— Temporal logics and formal models of action.
— System architectures and data stores for planning.

9. AN OVERVIEW OF THE EDINBURGH O-PLAN PROJECT

Finally, as an example of the approach being taken in the research community, the knowledge-based planning system O-Plan (Open Planning Architecture) currently being developed at the University of Edinburgh (Currie and Tate, 1985) will be outlined.

9.1 Planning and control

Computer-based systems for planning and control assume that a great deal of preparatory work has been done to describe the task and the detail of a method to carry out that task. Much of the information used in this preparation is then discarded. However, it is this same information that is of value during the control of the execution of a plan and for replanning when problems arise.

Earlier UK Science and Engineering Research Council (SERC) supported research ("Planning: a joint AI and OR approach") tackled the first issue of seeking to provide computer aids to prepare the initial plan. This involved providing a language ("task formalism") in which information about the separate activities possible in a planning application could be described. The NONLIN AI Planner (Tate, 1977) was designed to take such descriptions and a skeleton plan or task outlined at an abstract level and to choose appropriate activities to produce a detailed plan to perform the given task. The planner and domain description language were tested on an electricity turbine overhaul problem using data provided by the UK Central Electricity Generating Board (Daniel, 1983). The internal knowledge representation and plan explanantion and a plan execution monitoring framework could be based. However, it is only recently that these ideas have begun to be explored.

9.2 The research area

The Edinburgh research involves the development of techniques and the production of a prototype system to cover the full range of activities involved in a command, planning and control task such as project management. This involves:

— knowledge elicitation of information that describes the application area
— task or goal representation
— constructing a plan to carry out the desired task within the constraints set
— allowing user interaction, explanation and changes of base assumptions during planning
— interfacing to application dependent sub-systems which provide detailed planning knowledge (such as route planners based on maps)
— monitoring the execution of plans to ensure that they meet their required objectives
— allowing for recovery after failure of parts of the plan

We are principally concerned with the production of a planner that can act as part of an overall system for decision making and control in applications such as civil engineering, project management and batch manufacturing. However, the techniques to be developed are domain independent.

The research is anchored in a prototype computer-based planning system called "O-plan" (open planning architecture). The name reflects our aim of providing clear modular interfaces to allow experimentation and integration of work underway on related knowledge-based systems. Currie and Tate (1985) gives more details of the O-Plan system.

9.3 Present work — integrated planning systems

The starting points for the new O-Plan work were:

— The pre-existing NONLIN planner and extensions made by others to handle time and resource constraints
— The concept of Goal Structure to capture the 'intent' of plans
— The Task Formalism domain description language

We have begun the development of a prototype planner based on a simplification of the new architecture. This is being implemented in Prolog under UNIX on SUN workstations. The implementation is intended to aid our further understanding of the problems being tackled and help to define the later work programme.

9.4 O-Plan project work areas

Fig. 2 shows an overview of the O-Plan Architecture. The following sections indicate the scope of the work involved in developing suitable techniques and demonstrating them in the prototype.

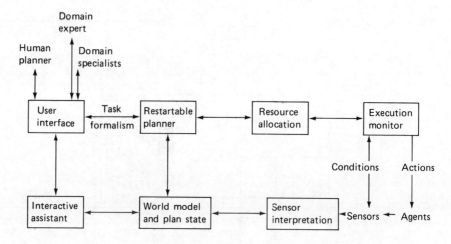

Fig. 2 — Overview of the O-Plan system.

9.4.1 *Knowledge representation*
The language used for domain description must allow for the capture of many types of heuristic information about the domain. It should admit incremental definition and give aids on which the specification of the various parts can be verified. A design suited to the provision of an intelligent user interface is necessary.

A design for the Task Formalism (TF) Language is now available which we believe is able to represent the domains of interest. As the O-Plan planning system develops we expect to be able to handle progressively more parts of this language.

Many of our concerns with the new planning architecture also relate to how a plan under construction should be represented and how dependencies in it should be maintained. Our research is exploring various fundamental knowledge representation issues of general concern in artificial intelligence and knowledge-based systems.

9.4.2 *Knowledge elicitation*
We have already designed and built a demonstration implementation of a user interface to describe a domain to the planning system. This is a graphically oriented workstation using PERT-type networks as the main information presentation aid on an ICL PERQ 2. A video has been made of the functions of the workstation. Fig. 3 shows sample screens. The workstation is intended to cope with all the types of user who play various roles with respect to a planning environment.

Consideration has been given to the imposition of a requirements analysis methodology into the user interface to impose structure on the knowledge elicitation process. The use of Systems Designers' CORE methodology has been considered (on an M.Sc. project) as this has much in

138 APPLICATIONS OF KNOWLEDGE-BASED PLANNING SYSTEMS [Ch. 11

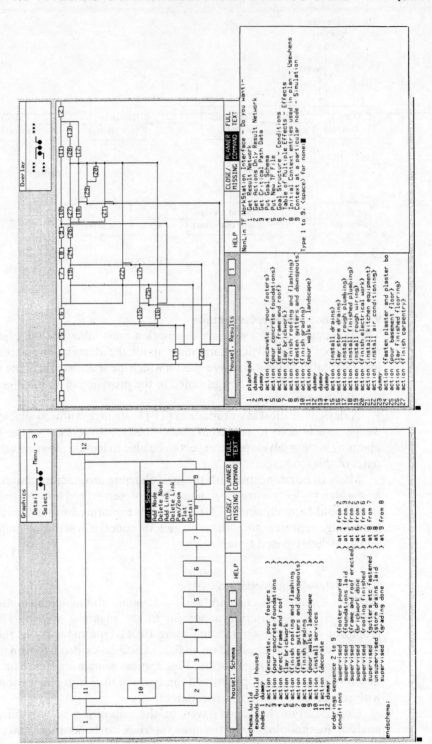

Fig. 3 — Sample TF workstation screens.

common with the structures used in the present Task Formalism. There is a clear overlap between knowledge elicitation for a planning domain and the specification of modules in a software engineering environment.

9.4.3 *Planner design*

The central element of our work is the development and demonstration of a new generation AI planner able to cope with much more realistic problems than earlier systems.

The O-Plan planner design is based on NONLIN and Goal Structure but is orientated around the gradual restriction of choices (by symbolic constraint propagation) and the opportunistic selection of the focus of attention. A blackboard and layered agenda control structure is used. At present, we are engaged in building a prototype planner which has most features of the O-Plan planner in a simplified form. However, in the full O-Plan system, the dependencies between decisions taken will be recorded symbolically to enable a single most promising solution to be maintained rather than keeping alternative partial solutions as is needed for decisions in the present prototype. Failures at planning or execution time will be analysed and the decision dependencies used to "repair" the partial plan and to continue on to a solution.

Also being considered in the planner design are:

— interface to the "Functions in Context" data model (see below)
— addition of user interaction, manual control, diagnostics and result presentation
— addition of planning with time and resource constraints and external event handling
— addition of intelligent interfaces to sub-planners, instruction compilers, body modellers, route finders or other computationally defined sources of information of expertise
— integration of the planner with an execution monitor so that it can accept back partially executed plans and failure conditions and continue to try to find a solution.

9.4.4 *Functions in context data model*

We have developed an entity/relationship data model with context varying values. This is an essential component of the planning system envisaged. The functions to be performed by this system are presently implemented in our prototype by a simple tuple store with incremental fully ordered changes of values (like many AI knowledge representation languages). On top of this, we represent a partially ordered set of changes of the values on entities. This scheme is a known performance bottleneck of the existing AI planners.

We wish to abstract out the full data model which is necessary to support the planner and to provide a clean interface to this as a separate system. This will facilitate the incorporation of an efficient software implementation or hardware content addressable memories in future. We believe that hardware assistance to support the data model will be necessary to provide knowledge based planning systems able to cope with realistic tasks. The data model will store clauses of the form:

$$f(p1, p2 \ldots) = \text{value at point}$$

and allows for retrieval of partially specified matching items (via a lazy evaluated "generator" and "try-next" mechanism). The system will allow for the specification and retention of contexts as a point in a partially ordered set of such points. Ambiguous inheritance of the value of patterns is possible within the context structure specified. The system (a planner in the O-Plan case) which uses the data model must interpret any ambiguity in an application dependent way. The "Functions in Context" data model has very general applicability and could be used in areas such as software engineering, mutliple-inheritance tape systems, etc. as well as in AI planning.

We have already published a number of working papers and discussed our ideas with UK workers engaged in building hardware large-scale knowledge base machines (at Manchester and Strathclyde). The O-Plan research expects to verify the utility of the data model using one of the hardware knowledge bases being constructed in the UK. This would also serve as a serious benchmark for the hardware in a situation where the data is highly dynamic and heavily qualified.

9.4.5 *Plan execution, monitoring and re-planning*

Goal Structure is a high level representation of information about a plan which states the relationship between the individual actions in the plan and their purposes with respect to the goals of sub-goals they achieve. This information is used by the planner to detect and correct conflicts when higher level plans are refined to greater levels of detail.

Goal Structure can be extended to represent information on which an execution monitor can operate effectively. The Goal Structure statements represent precisely the outcome of any operation which should be monitored. If lower level failures can be detected and corrected while preserving the stated Goal Structure, the fault need not be reported to a higher level. The implications of any propagated failure are computable and corrective action can be planned.

The techniques are equally applicable to the generation of plans and the monitoring of their execution for fully automated or fully manual situations as well as the more usual mixed environment. Hence we are suggesting an overall organisation within which individual modules for planning or control can operate.

Tate (1984) describes the proposed execution monitoring framework and discussed this with groups in the UK engaged in building intelligent sensors and local recovery sub-systems for flexible manufacturing systems (UK National Engineering Laboratory and Aberystwyth). The O-Plan research will develop a computer based execution monitor and interfaces to the AI planner to cope with high level re-planning after serious failures. This will be used to verify our ideas and to further explore the interfaces to realistic sensors and effectors.

9.4.6 Applications of AI planning

O-Plan will be tested via the development of domain descriptions in the enhanced task Formalism for various applications. The automatic generation of plans for representative tasks would then be undertaken. The applications of interest to us at present include:

— small batch manufacturing and assembly systems
— logistics, command and control
— project management (such as in civil engineering and software production)
— spacecraft mission sequencing

The Voyager spacecraft mission sequencing example will use the domain description given to the NASA Jet Propulsion Laboratory DEVISER planner (itself based on our own earlier NONLIN system). This will provide a realistic trial and comparative data.

We are also engaged in the UK Alvey Programme's PLANIT Community Club project to show the benefits of manipulating pre-existing plans in the "knowledge rich" form that would have been generated by an AI planner. This will provide detailed information on three domains:

— back axle assembly at Jaguar Cars
— a software production project at Price Waterhouse Associates
— a job shop problem at Harwell Atomic Energy Authority

The same projects will be of value in providing realistic data on which to try out the automatic production of plans in these domains (a development assumed by the Alvey PLANIT Community Club project).

This part of the project will also be concerned with the specification (in conjunction with others) of suitable interfaces to an AI planning system from higher level management, design and aggregate (or capacity-based rough-cut) scheduling software and to lower level detail planners, instruction generators, resource smoothing packages, intelligent sensors and robotic devices.

ACKNOWLEDGEMENTS

This work is funded by the UK Science and Engineering Research Council and through the Alvey Directorate. The support of Systems Designers is gratefully acknowledged.

SHORT PLANNING BIBLIOGRAPHY

[1] Allen, J. F. and Koomen, J. A. (1983) Planning Using a Temporal World Model, IJCAI-83, pp. 741–747 Karlsruhe, West Germany. [TIMELOGIC]

[2] Currie, K. and Tate, A. (1985) O-Plan: Control in the Open Planning Architecture Expert Systems 85, BCS Workshop Series, Cambridge University Press, Proceedings of the BCS Expert Systems Group Conference, Warwick, UK, December 1985 [O-PLAN].

[3] Daniel, L. (1983) Planning and Operations Research, in *Artificial Intelligence: Tools, Techniques and Applications*, Harper and Row, New York. [NONLIN + Decision Graph].

[4] Doran, J. E. and Michie, D. (1966) Experiments with the Graph Traverser Program, *Proceedings of the Royal Society, A*, pp. 235–259. [GRAPH TRVERSER].

[5] Doyle, J. (1979) A Truth Maintenance System, *Artificial Intelligence*, **12**, pp. 231–272.

[6] Erman, L. D., Hayes-Roth, F., Lesser, V. R., and Reddy, D. R. (1980) The HEARSAY-II Speech-understanding System: Integrating Knowledge to Resolve Uncertainty, *ACM Computing Surveys*, **12**, No. 2. [HEARSAY-II]

[7] Fikes, R. E. (1982) A Commitment-based Framework for Describing Informal Cooperative Work, *Cognitive Science*, **6**, pp. 331–347.

[8] Fikes, R. E. and Nilsson, N. J. (1971) STRIPS: a New Approach to the Application of Theorem Proving to Problem Solving, *Artificial Intelligence*, **2**, pp. 189–208. [STRIPS]

[9] Fikes, R. E., Hart, P. E. and Nilsson, N. J. (1972a) Learning and Executing Generalised Robot Plans, *Artificial Intelligence*, **3**, [STRIPS/PLANEX]

[10] Fikes, R. E., Hart, P. E. and Nilsson, N. J. (1972b) Some New Directions in Robot Problem Solving, in *Machine Intelligence 7*, Meltzer, B. and Michie, D., eds., Edinburgh University Press. [STRIPS]

[11] Fox, M. S., Allen, B. and Strohm, G. (1981) Job Shop Scheduling: an Investigation in Constraint-based Reasoning, IJCAI-81, Vancouver, British Columbia, Canada, Canada, August 1981. [ISIS-II]

[12] Hayes, P. J. (1975) A Representation for Robot Plans, IJCAI-75, Tbilisi, USSR, September 1975.

[13] Hayes-Roth, B. and Hayes-Roth, F. (1979) A Cognitive Model of Planning, *Cognitive Science*, pp. 275–310. [OPM]

[14] McDermott, D. V. (1978) Planning and Acting, *Cognitive Science*, **2**.

[15] Miller, D., Firby, J. F. and Dean, T. Deadlines, Travel Time and

Robot Problem Solving, IJCAI-85, Los Angeles, USA. August 1985. [FORBIN]

[16] Newell, A. and Simon, H. A. (1963) GPS: a Program that Simulates Human Thought, in Feigenbaum, E. A. and Feldman, J. eds., *Computers and Thought* (McGraw-Hill, New York, 1963). [GPS]

[17] Rich, C. (1981) A Formal Representation for Plans in the Programmer's Apprentice, IJCAI-81, pp. 1044–1052, Vancouver, British Columbia, Canada.

[18] Sacerdoti, E. D. (1973) Planning in a Hierarchy of Abstraction Spaces, Advance papers of IJCAI-73, Palo Alto, Ca., USA. [ABSTRIPS]

[19] Sacerdoti, E. D. (1977) *A Structure for Plans and Behaviour*, Elsevier-North Holland. [NOAH]

[20] Schank, R. C. and Abelson, R. P. (1977) *Scripts, Plans, Goals and Understanding*, Lawrence Erlbaum Press, Hillsdale, New Jersey, USA.

[21] Siklossy, L. and Dreussi, J. (1975) An Efficient Robot Planner that Generates its own Procedures, IJCAI-73 Palo Alto, Ca., USA. [LAWALY]

[22] Stallman, R. M. and Sussman, G. J. (1977) Forward Reasoning and Dependency Directed Backtracking, *Artificial Intelligence*, **9**, pp. 135–196.

[23] Stefik, M. J. (1981a) Planning with Constraints, *Artifical Intelligence*, **16**, pp. 111–140. [MOLGEN]

[24] Stefik, M. J. (1981b) Planning and Meta-planning, *Artificial Intelligence*, **16**, pp. 141–169. [MOLGEN]

[25] Sussman, G. A. (1973) A Computational Model of Skill Acquisition, M.I.T. AI Lab. Memo no. AI–TR-297. [HACKER]

[26] Tate, A. (1975) Interacting Goals and Their Use, IJCAI-75, pp. 215–218, Tbilisi, USSR. [INTERPLAN]

[27] Tate, A. (1977) Generating Project Networks, IJCAI-77, Boston, Ma., USA. [NONLIN]

[28] Tate, A. (1984) Planning and Condition Monitoring in an FMS, Proceedings of the International conference of Flexible Automation Systems, pp. 62–69, Institute of Electrical Engineers, London, UK. July 1984. [NON-LIN]

[29] Tate, A. (1985) A Review of Knowledge based Planning Techniques, Expert Systems 85, BCS Workshop Series, Cambridge University Press, Proceedings of the BCS Expert Systems Group Conference, Warwick, UK, December 1985.

[30] Vere, S. (1983) Planning in Time: Windows and Durations for Activities and Goals, *IEEE Trans. on Pattern Analysis and Machine Intelligence*, **PAMI-5**, No. 3, pp. 246–267, May 1983. [DEVISER]

[31] Waldinger, R. (1975) Achieving Several Goals Simultaneously, SRI AI Center Technical Note 107, SRI, Menlo Park, Ca., USA.

[32] Warren, D. H. D. (1974) WARPLAN: a System for Generating Plans, Dept. of Computational Logic Memo 76. Artificial Intelligence, Edinburgh University. [WARPLAN]

[33] Warren, D. H. D. (1976) Generating Conditional Plans and Programs, Proceedings of the ASIB Summer Conference, pp. 344–354, University of Edinburgh, UK, July 1976. [WARPLAN-C]
[34] Wilensky, R. (1978) Understanding Goal-based Stories, Dept. of Computer Science, Yale University, Research Report No. 140.
[35] Wilensky, R. (1981) Meta-planning: Representing and Using Knowledge about Planning in Problem Solving and Natural Language Understanding, *Cognitive Science*, **5**, pp. 197–233.
[36] Wilensky, R. (1983) *Planning and Understanding*, Addison-Wesley, Reading, Mass.
[37] Wilkins, D. E. (1983) Representation in a Domain-Independent Planner, IJACAI-83, pp. 733–740, Karlsrhue, West Germany. [SIPE]

12

A Prototype Expert System for Configuring Technical Systems

Michael Vitins, Brown Boveri Research Centre, Artificial Intelligence Group, CH-5405 Baden, Switzerland

1. INTRODUCTION

The key underlying idea of a knowledge-based system is to describe a specific domain of thinking by means of a suitable representation language. This knowledge can then be made available to improve and accelerate the process of solving problems in that specific knowledge domain. This paper describes an expert system shell that has been designed to exploit the use of knowledge representation techniques in the area of system configuration.

The main advantage of the knowledge-based approach is that the engineers responsible for the expert system will be able to develop, comprehend and maintain the knowledge base. In the conventional software approach, a program for each specific application domain has to be specified, designed and maintained with the support of programmers, thus incurring high costs that can be avoided in the knowledge-based approach.

The first successful and best known expert system for configuring systems was R1 of DEC [7], that was later further developed to XCON. In R1, all domain knowledge is defined in terms of pattern-action decision rules, called production rules.

As the number of rules in the knowledge base becomes large, increasing effort must be invested to manage the stucture of the rules and to select the active rules in a given context [6]. The most well known and effective representational form for structural and behavioral descriptions currently in use are frames, originaly inspired by Minsky [9]. The first frame implementations focussed on structural descriptions, e.g. FRL [10,11], KL-ONE [3] and KRL [2], in a manner that could easily be comprehended by an end user. The frame concept was shaped into a popular Lisp programming concept by Winston [15].

The integration of frames and rules to form an overall system that provides structural descriptions and the control of reasoning has been a recent achievement (e.g. LOOPS [14], KEE [5], CENTAUR [1], and DESIGN [8]). These systems have demonstrated the ease in which taxonomy descriptions, value restrictions, behavioural properties and reasoning services can be incorporated into a single system. The great power and potential of such a system is due to the fact that the search space for the reasoning process can be selectively controlled (and reduced) by exploiting the structure of the domain knowledge.

The emphasis of this study is to apply and experiment with frame-based AI representation and reasoning techniques to the problem of configuring large and complex technical systems. The frame methodology was selected for its intrinsic ability to support the organization and storage of knowledge, which was considered to be a major characteristic in this domain. Alternative approaches are the use of logic or production systems, These two approaches, however, do not provide much support for structure descriptions. Logic is advantageous when the application domain can be axiomatized while production systems are well suited when most of the domain knowledge can be expressed in the form of situation-action heuristics and judgement (that is, if–then rules).

The implementation language chosen for this study was the AI language Common Lisp. Common Lisp is a Lisp dialect that meets industrial standards and is becoming widely available. It was designed to eliminate the problems that arise due to the presence of a large number of Lisp dialects and to specifically meet the following goals in the domain of Lisp programming: commonality, portability, consistency, expressiveness, compatibility, efficiency, power, and stability, as described in Steele [13].

In order to emphasize the experimental nature of the study and to keep the program behavior under control, a decision was taken to not use any of the existing and commercially available frame-based packages as a baseline for further development. This approach encourages the development of concepts that are particularly suitable for the configuration domain. On the other hand, the decision has the disadvantage that some of the well known frame concepts had to be reimplemented. Clearly, this disadvantage is only of the short term nature. In later versions, when large and complex real world applications will be dealt with, additional implementation problems and optimizations will likely need to be considered. The system currently runs on a VAX 11/780 and work is under way to port the program to an IBM PC AT.

2. THE CONFIGURATION PROBLEM

Typical examples of the sort of technical systems considered in this paper deal with electrical generation, transmission, and distribution systems, such as thermal power plants, power control systems, and switchgear stations.

The two following activities are essential in order to configure a technical system:

— defining the original problem and the functions that the technical system must provide;

— defining the structure and components of a solution that will meet the specified problem.

The knowledge required to carry out the first activity must capture the notions relevant to the customer world, that is the party buying the technical system. The second activity is based on the knowledge of the vendor of the technical system and it describes the possible structures and components needed to actually build the technical system to achieve the specified goals.

The task of a project engineer, configuring a specific system, is to use such knowledge to express the customer problem and to find a solution for that problem. This process is shown schematically in Fig. 1.

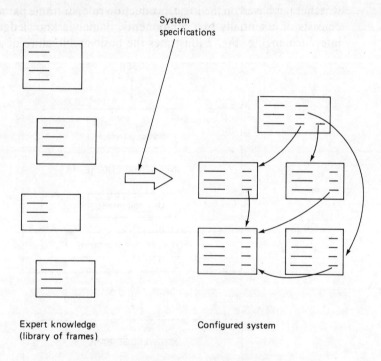

Fig. 1 — Configuring a technical system.

We assume that most of the expert knowledge is expressed in terms of frames (see Section 4). It is the task of a knowledge engineer to prepare a library of frames that are suitable for representing the objects in the problem and solution domains. Typically, there will be frames that describe technical functions, physical devices, checklists containing subproblems to be solved, and output forms for communicating specific information to the many people involved in the configuration process.

Although each individual technical system is unique, the objects occurring in both the problem and solution domains, are generally quite stable in the sense that they are time-tested and do not change dramatically over time. Due to the lack of sufficient support tools to explicitly express and manipulate domain knowledge, we find a rather high amount of redevelopment whenever new problems and solutions are considered. A major promise of the knowledge-based approach is to make knowledge explicit, thus emphasizing commonality and reusability.

Another major feature that a configuration support system must offer is a high degree of user interaction and flexibility. It must be possible to easily modify both the frames in the library and the frames describing the current state of the configured technical system.

3. ARCHITECTURE OF A KNOWLEDGE-BASED EXPERT SYSTEM

A distinguishing architectural feature of an expert system, irregardless whether it is based on the logic, production rule, or frame paradigm, is that it consists of essentially two components, namely a knowledge base and an inference engine. Fig. 2 illustrates the basic architecture of a frame-based

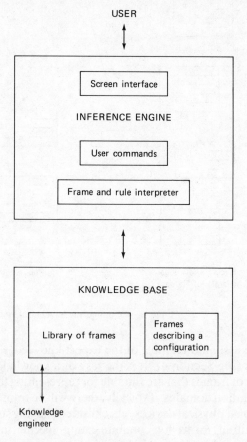

Fig. 2 — Basic architecture of a knowledge-based expert system.

expert system, as applied to the configuration problem.

The user of the expert system is expected to be a project engineer or a sales person, who may or may not possess the complete technical expertise of the application domain. The primary goal of the user is to configure a technical system that will satisfy customer specifications and that can be implemented as far as possible in terms of the predefined system components given in the frame library. The purpose of the inference engine is to guide the user in entering and accessing data and to assist the user in controlling the process of configurating the technical system in a user-oriented way.

When the user initiates the expert system the first time, the knowledge base will generally consist solely of the library of frames that have been previously designed for the domain of application at hand by a knowledge engineer. As user input is received by the system, the inference engine will activate or instantiate the appropriate knowledge to accumulate new data that contributes to the description of the resulting technical system.

The formalization and codification of the domain expertise, which is the task of a knowledge engineer, is a highly critical activity. The choice of the library frames has a fundamental impact on the feasibility and success of the expert system. For this reason, it is important that the library frames can be modified and improved by the knowledge engineer both before and during the configuration of a specific system in order to incorporate new experiences and new facts immediately as they become available.

4. KNOWLEDGE BASE

A frame is a three-level hierarchical data structure that can be used to store information about a particular object or aspect of a system.

A frame can have any number of slots. Each slot names an attribute of the object at hand.

Each slot can have any number of facets. A facet specifies an attribute of the associated slot, and thus provides a means to structure and classify that slot. Facets prove to be very useful not only for organizing information about slots, but also for controlling the behaviour of the expert system, as will be described in Section 5.

Finally, a facet can have any number of values. A value is typically given in the form of a number or a symbol representing the name of an atomic object, a frame, a (Lisp) function, or a rule.

Fig. 3 illustrates an example of the use of frames for describing frame instances and frame types (that is, classes of frames) in the domain of medium voltage switchgear yards (i.e. substations).

The frames in Fig. 3 are to be interpreted as follows. The frame called Niederbipp is of the type Substation, has the nominal voltage 36, the nominal current 800, the construction metal clad, and the fields F1 and F2. F1 is a field type frame, that is a component of Niederbipp and consists of the apparatus F1-Breaker and F1-Ct, which happen to be the names or further frames. The Substation frame is a type frame which represents the class of all substations. The facets of this frame will be described in Section 5.

In the example shown in Fig. 3, the frames Niederbipp, F1, and F2 represent frame instances that have been created by a project engineer to

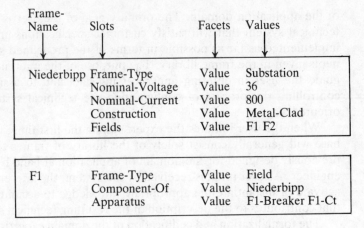

(a) Frames describing components of a configuration

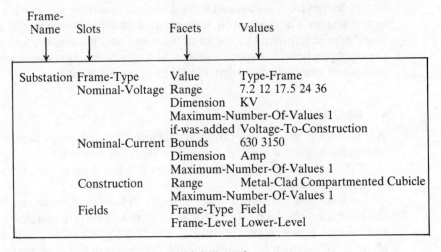

(b) A library frame

Fig. 3 — Substation Frames.

describe a particular configuration. The frame Substation, on the other hand, represents a type (or class) frame and has been created by a knowledge engineer to express all relevant properties of a substation.

5. INFERENCE ENGINE

A number of reserved names for slots, facets and values contain the key knowledge that controls the reasoning of the inference engine. The reserved names that are employed by the inference engine are independent of the application domain.

The knowledge engineer is free to choose the names of the frames, slots, facets, and values in order to suit the application at hand, but he must use the reserved names in accordance with the intentions of the inference procedures.

Frame-based representations explicitly support the following three types of reasonings procedures:

— Inheritance
— Value constraints
— Procedural attachment.

5.1 Inheritance

Inheritance allows information to be shared among multiple frames. The reasoning about a large body of knowledge can thus be concentrated in dedicated descriptions within appropriate frames. Inheritance supports the access of data in the knowledge base in a controlled manner, avoiding the use of multiple specifications and avoiding general purpose and somewhat inefficient deduction mechanisms, such as logic theorem proving or production systems, Fikes [4].

Inheritance is effectively "built in" in the frame machinery and comes in many forms. The forms of inheritance can be derived from the taxonomies that reflect the relationships between the frames. The following types of links between frames are particularly relevant and can be used to provide inheritance:

— The Member Link:

A frame instance can be defined to be a member of some type frame. This means, that the frame instance can inherit all the slots, facets, and values from the type frame.

In Fig. 3, for example, Niederbipp is a frame instance of the type Substation. All data that is specific for Niederbipp is stored in the Niederbipp frame, whereas other informaton about Niederbipp, that is common to all substations, is stored in the Substation frame.

It is possible to suppress the inheritance of some specific piece of data from the type frame by explicity deleting that specific data link from the instance frame. It is also possible to add further slots and facets to the instance frame that are not present in the type frame. In the extreme, an instance frame can be modified so drastically that it has very little — if not nothing at all — in common with its type frame. In such cases, it would probably be worth while to consider introducing a new independent type frame, in order to make that structure available for other similar instances.

— The Component Link:

A configuration can often be viewed as consisting of a hierarchy of components. Each component, being represented as an instance frame, is described in terms of further components, until finally atomic components are reached. The component hierarchy allows a component to inherit values, that are not explicitly stored in that component, from its parent component. Thus, if some specific data cannot be found immediately in a component instance frame (or in its type frame), the frame named in the component-of slot is used to search for that data. In its simplest form, the search stops as soon as a first value has been found.

In Fig. 3, the frame F1 is seen to be a component of the frame Niederbipp. Thus the nominal voltage of F1 can be derived to be equal to 36, thanks to the component inheritance.

— The Class Link:

A type frame can inherit its data from yet other frames. For example, the type frame Substation could be considered as being a subclass of an electrical plant. This sort of inheritance supports the modelling of fairly realistic and sophisticated views of the expert's world, but also introduces additional complexities due to the possibility of intermixing type and instance frames. The current version of the program does not support the use of the class link.

Further types of links between frames can be added — as needed — by the knowledge engineer in order to make domain-dependent relations explicit, such as functional dependencies or physical connections between objects.

5.2 Value constraints

Facets can be used to restrict the possible values that the value facet of a slot can possess.

If a facet called range is provided, the inference engine will accept values to that slot only if the value coincides with one of the values specified in the range facet. Further, a facet called bounds can be used to store lower and uper bounds for the permissible values of a slot. A facet called maximum-number-of-values will ensure that the number of values of a slot will not exceed the specified number. If any of these constrtaints are not satisfied, the system will ignore the new value to be inserted and will issue a warning message on the screen.

The facets frame-level and frame-type contain information concerning the frames stored in the slot at hand. The frame-level facet indicates whether the frames stored in the slot are on a lower, higher, or the same hierarchical component level as the frame at hand. The frame-type facet provides the names of the possible types of the frames that can be stored in the slot at hand. These facets are automatically referred to and employed when a new value is added to the slot at hand.

5.3 Procedural attachment

An effective technique to deal with the behavioral representation of complex structures is to make procedural knowledge accessible locally within the frames. Thus, inference related knowledge can be found very efficiently without search when a frame is being accessed or modified. On the other hand, locally stored inference related knowledge is no longer arbitrarily accessible, hence higher demands must be posed on the compromise between the generality and speciality of the domain representation.

A special case of procedural attachment is given by the value constraints mentioned above. Each value constraint is an indication for the inference engine to behave in a predefined procedural way when a value is deposited in a slot.

A more general approach to procedural attachment is to store the procedures for manipulating and computing values within predefined facets and to arrange the inference engine to evaluate those procedures whenever specific situations arise. Functions that are evaluated when a value is added, changed or removed are called DEMON FUNCTIONS. The following facet names have been implemented to store the names of demon functions: if-to-be-added, if-was-added, if-needed, and if-removed, following the spirit of Winston [15]. These facets allow demon functions to be triggered selectively, before a value is added to the slot, after having added a value to the slot, to compute a value on an if-needed basis, or to trigger actions after a value was deleted. If a demon stored in an if-to-be-added facet returns nil, that value will be rejected.

A demon function can be defined either as a Lisp function or in the form of an if-then rule or — in future — in any other user oriented form for which an interpreter can be written. Rules have the general form:

IF premise THEN action,

where premise generally involves testing the values in one or more slots and action generally involves inserting one or more values into slots.

An example of a simple rule is

IF (nominal-voltage = 36) THEN (construction = metal-clad),

which happens to be the definition of the rule named voltage-to-construction that was referred to in Fig. 3(b). When a rule is invoked from within a frame, the slot names (i.e. nominal-voltage and construction) occurring in the rule refer to those of that particular frame.

The basic mode of applying these rules is FORWARD-CHAINING. As soon as the knowledge base has received a new slot value, say due to a user input, the inference engine invokes all the rules (and other demon functions) given in the if-was-added facet, thus attempting to derive new conclusions. If a new conclusion could be reached, that conclusion is deposited into some appropriate slot by the inference engine. That conclusion can invoke further rules, which may lead to still more conclusions. A simple user input can thus possibly generate a whole avalanche of new conclusions.

Not surprisingly, a rule is internally represented as a frame instance of the type rule, that consists of at least two slots, one for storing the premise and the other for the action.

The slotwise storage of demon functions via facets can sometimes lead to a repetition of a group of functions that occurs collectively over and over again. In order to avoid such repetition we allow facets to contain not only demon functions but also frames that can be used to access further demon functions. Those frames again contain other demon functions and frames for accesing still further demon functions. The resulting mechanism for storing demon functions is thus recursive.

No provision is made to incorporate uncertaincy with the values entered into the frames, in contrast to several rule-based expert systems, such as MYCIN [12]. It was felt that uncertaincy values themselves are more a source of confusion than help for a user.

The values that are stored in the knowledge base are considered to be valid. However, they may be altered at a later time as more detailed and reliable data becomes available, unless value constraints prohibit this. The ability to alter previously stored data, may lead to serious consistency problems, if used incarefully, because it may be difficult to undo the result of previous rule invocations.

Ideally, all domain dependent procedural knowledge should be formulated in a manner that is declarative and comprehensible for the user (and for the inference engine), hence, the rule notation should be favoured.

It is usually not necessary to make ALL procedural and algorithmic knowledge declarative and explicitly comprehensible to a user. An example of procedural knowledge that is of little interest to the average user is the representation of complex but well established specialised expertise, that is highly unlikely to change. Another example is expertise that the user should not be allowed to see or modify. In such cases, the language Lisp itself may be the appropriate representation language, since the vast possibilities of Common Lisp and its programming environment are available directly. We recall that a Lisp function can be invoked by the inference machine via one of the demon facets. Another way to use a Lisp function is to build it into the premise or action side of the rule, where it can be evaluated via rule invocation.

A lot of research has been and is currently underway to incorporate additional representational mechanisms, for example multiple inheritance paths, deep reasoning, and handling informal requests. Such proposals provide more ambitious and powerful capabilities but are on the other hand also more complex and possibly less easy to use. Further developments are planned in order to experiment with such concepts to identify their strengths and weaknesses for the configuration problem.

6. USER COMMANDS

This section gives an overview of some of the commands that are at the disposal of the program user and the knowledge engineer.

The user can display any frame instance on the screen in a semi-graphic form either by specifying a frame name or by using the currently cursored value that is present on the screen. Commands are available for moving the cursor position within the frame and for changing the section of the frame that is visible on the screen.

When entering values into a slot, the constraints that must be obeyed for that slot are displayed in a special comment window, before input data can be accepted. The user can ask the system for guidance, that is for the next slot that has not yet been given a value. An explain command displays a list of the demon activities since the last major user input. New frame instances can be created from scratch or by copying previously existing instance frames, either with or without modifying slot values.

A number of commands are available to inform the user about the contents of the knowledge base and the state of the configuration. The configuration can be searched exhaustively for specific data, concerning frames, slots, facets, and values by means of a pattern matcher.

A number of miscellaneous commands are provided to support the overall use of the expert system. All user commands can be found by a help command and are documented on the system. Commands for loading, saving and compiling files that contain domain specific knowledge are provided. We recall that all domain knowledge is represented as frames and Lisp functions. The frames stored in non-compiled files are available in a user-oriented format that is independent of the internally used Lisp representation.

The user can modify the slots, facets, and values of instance frames in a practically limitless way. Manipulations of type frames can only be carried out by privileged users, typically by knowledge engineers. A simple frame editor is provided for creating and modifying type frames.

7. IMPLEMENTATION

The concepts presented in the previous sections have been implemented in the form of an interactive software package, written in the programming language Common Lisp, executing on a VAX 11/780. The program makes use of windowing techniques, as provided by the VAX/VMS Screen Management Guidelines (SMG) run-time library routines.

The main program loop accepts user input and triggers the appropriate internal user commands. The inference engine currently consists of 14 packages, each having an average of 583 lines of source code (including documentation strings and comments). A package is a data structure that provides a separate name space for its symbols; hence, dependencies between different packages can be controlled via import/export relations. A package is a structure suitable for implementing an abstract data type, that is a collection of functions operating on some internal and private data structure.

8. EXAMPLE DIALOG

The purpose of this section is merely to illustrate a typical user dialog by means of a very simple example, rather than to present a particularly impressive or intelligent real life example.

Fig. 4 illustrates the user view of an instance frame called Niederbipp,

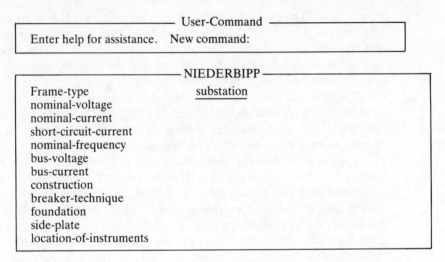

Fig. 4 — The frame Niederbipp.

shortly after it was defined to be of the type Substation. Note that the screen displays the slots relevant for substations, as defined by the substation frame. The user is only shown the contents of the value facet of the slots, since it is this information that essentially defines the configuration. Additional user commands must be invoked for displaying the contents of other facets.

Fig. 5 shows a snapshot of the screen in the course of defining the nominal voltage to become equal to 36.

Fig. 6 shows the screen after the nominal voltage was set to 36 and after the explain command was entered. Note that the slot called construction was automatically set to metal-clad, as a consequence of the invocation of the rule voltage-to-construction that was already mentioned in Section 5 and shown in the if-was-added facet in Fig. 3(b).

9. CONCLUSIONS

Frames have been demonstrated to provide a powerful and economic representation technology. This paper indicates that even the simplest realizations of inheritance, value constraints, and procedural attachment

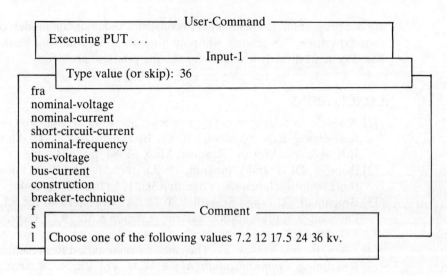

Fig. 5 — Putting a value into a slot.

Fig. 6 — The explanation of rule invocation.

provide sufficient power to be of substantial value for configuring technical systems.

One of the advantges of the frame approach is that domain experts who are not programmers are enabled to construct and maintain models of their own expertise. This feature will play a major role in making knowledge-based systems directly available for solving practical problems

CONCLUSIONS

[1] Aikins, J. S.: A representation scheme using both frames and rules. In *Rule-Based Expert-Systems*, B. G. Buchanan and E. H. Shortliffe, Eds. Addison-Wesley, Reading, Mass., 1984, pp. 424–440.
[2] Bobrow, D. G. and Winograd, T.: An overview of KRL, a knowledge representation language. *Cognitive Sci.* **1**, 1 (Jan. 1977), 3–46.
[3] Brachman, R. J. and Schmolze, J. G.: An overvierw of the KL-ONE knowledge representation system. *Cognitive Sci.* **9**, 2 (Apr. 1985), 171–216.
[4] Fikes, R. and Kehler, T.: The role of Frame-based Representation in Reasoning, *Communications of the ACM*, Vol. 28, No. 9, Sept. 1985.
[5] Kehler, T. P. and Clemenson, G. D.: An application development system for expert systems. Syst. Softw. **3**, 1 (Jan. 1984), 212–224.
[6] Hayes-Roth, F.: Rule-Based Systems, *Communications of the ACM*, Vol. 28, No. 9, Sept. 1985.
[7] McDermott, J.: A rule-based configurer of computer systems, April 1980, Carnegie-Mellon, CMU-CS-80-119.
[8] Miller, A. J.: DESIGN — A Prototype Expert System for Design Oriented Problem Solving. *The Australian Computer Journal*, Vol. 17, No. 1, Feb. 1985.
[9] Minsky, M.: A framework for representing knowledge. In *The Psychology of Computer Vision*, P. Winston, Ed. McGraw-Hill, New York, 1975, pp. 211–277.
[10] Roberts, R. B. and Goldstein, I. P.: The FRL primer, Memo No. 408, Artificial Inteligence Laboratory, MIT, Cambridge, Massachusetts, 1977a.
[11] Roberts, R. B. and Goldstein, I. P.: The FRL primer, Memo No. 409, Artificial Inteligence Laboratory, MIT, Cambridge, Massachusetts, 1977b.
[12] Shortliffe, E. H.: *Computer-based Medical Consultations: MYCIN*. New York, American Elsevier, 1976.
[13] Steele, G. L.: *Common Lisp, The Language*, Digital Press, 1984.
[14] Stefik, M., Bobrow, D. G., Mittal, S., and Conway, L.: Knowledge programming in LOOPS: Report on an experimental course. *Artif. Intell.* **4**, 3 (Fall 1983), 3–14.
[15] Winston, P. H. and Horn, B. K. P.: *LISP*, 2nd Edition, Addison-Wesley publishing Company, 1984

Index

access methods (AM), 26
analog electronic circuits, 57ff, 118ff
analogical reasoning, 18
annotation, 125
apes, 69ff
   applications, 72
   distribution, 71
Aquarius
   architecture, 37
   Project, 36
artificial intelligence, 7, 12, 111
assembling/disassembling (of Prolog terms), 107
assertion, 72ff
atom, 59
attributed grammar, 92ff

backtracking, 46
   naive, 54
   semi-intelligent, 45, 54
backward chaining, 63, 64
Berkeley PLM, 36, 39
block, 59

causality
   path, 20
   relationship, 123
chemical analysis, 78
circuit, 20
   description, 58ff
   schema, 119
class, 124
   link, 152
classification, 78
clause, 41, 72ff
communication, 101
composite objects, 125
CONAD, 111ff
configuration

banking system, 115
   technical systems, 145ff
configuration adviser, 111ff
constant, 73
constraints, 134, 152
consumable resources, 134
context data model, 139

data base
   management system (DBMS), 23
   Prolog coupling, 23
   relational, 24ff
data manipulation language (DML), 23
data-dependency analysis, 40
declarative programming, 12, 74, 75, 99
DEDALE, 57ff
deduction, 74
deductive
   component, 24
   model, 83
default, 62, 81, 125
demon (function), 126, 153
dependency graph, 43
dependency-directed search, 133
description language (for circuits), 67
design, 117ff, 121
diagnosis, 20, 57ff, 117ff, 123
   process, 62
dialogue, 81, 156
distance transform, 91
dynamic
   alteration, 84
   assertion, 86
   deletion, 86

Educe, 23
efficiency (of logic programs), 78
electrical parameters, 62
entity/relationship model, 139

exoskeleton, 92
experience rules, 60
expert system, 12, 17ff, 57ff, 111ff, 145ff
    (see also knowledge-based system)
    empirical approach, 19
    first generation, 57ff
    shell (tool), 12, 57ff, 69ff, 111ff
    tool, 98ff
explanation, 18, 75, 157
external data base (EDB), 24

facet, 124, 149
fact, 72ff
factual knowledge, 58
fast prototyping, 13
forall construct, 78
forward chaining, 63, 64, 153
frame, 58, 103, 124
    Lisp realization, 148ff
    Prolog realization, 57ff, 117ff, 127
    link, 124
frame-based knowledge programming, 118
frequency modulator, 120, 123
function block diagram, 119
functional representation (description), 59, 119
functioning, 61

generalization, 42
generic objects, 58
goal, 41ff, 73
goal structure, 134, 140
Grid File, 109

heuristic knowledge, 57ff, 60
heuristic rules, 60
heuristic search, 133
hierarchical planning, 133
hierarchy
    frames, 124
    modules, 121
Horn clause logic programs, 72ff
human–computer interaction, 109
hypothesis, 61

if-then rules, 17, 19
image, 92
    primitives, 91ff
incremental programming, 13, 100
inference procedure (engine), 11, 64, 148, 150
inferential knowledge, 60
influence analysis, 62
inheritance, 124, 151
instantiation, 62
integration (Prolog and databases), 23ff
interacting goals, 133
interaction, 18
interactive programming, 13, 75, 100
interface, 139

justification, 18

knowledge
    elicitation, 137
    engineer, 70, 147
    engineering environment, 14
    representation, 14, 18, 101, 103, 137
knowledge-base, 11, 148
    dynamic alteration, 84
knowledge-based design, 117ff
knowledge-based diagnosis, 105, 117ff
knowledge-based management system (KBMS), 23
knowledge-based planning, 130ff
knowledge-based system (see also expert system), 7, 11, 17ff, 98ff, 111ff, 145ff

level (abstraction), 20
link, 59
    class, 151
    component, 151
    member, 151
LIPS (logical inference per second), 37
LISP, 146
logic programming, 72ff
    databases, 23
logic-based tools, 69ff

map coloring, 54
member (frame), 124
meta-rule, 19, 65
model description, 92
model guided interpretation, 91ff
model-based reasoning, 19
modelling, 119
modifiability, 77
Modula—Prolog, 98ff
    screen, 107
modula-2, 99, 101
    predicate, 108
module, 20, 120

negation, 73
negation-by-failure, 73, 123
node, 59

objects as resources, 134
operators (in Prolog), 102
O-Plan, 134, 135, 137
OSSI, 109

parallelism, 36
    AND-parallelism, 37, 43
    OR-parallelism, 37, 48
    unification parallelism, 37, 48
partially-ordered actions, 133

performance, 53
plan
   execution, 140
   monitoring, 140
planner design, 139
planning, 130ff
   applications, 131, 141
   priorities, 134
   problem areas, 134
   systems, 131
   taxonomy, 131
   techniques, 133
port, 120
priority analysis, 62
problem solving, 14
procedural attachment, 60, 93, 125, 153
procedural interpretation, 74
procedural knowledge, 62
procedural programming, 99
procedure splitting, 50
production system, 17
program development, 75
programmed logic machine PLM, 36
programming environment, 13, 98ff
programming style, 99
project engineer, 147
Prolog, 23, 36ff, 57ff, 69ff, 74ff, 98ff, 117ff
   benchmarks, 37
   database coupling, 23
   extended, 31
   frame system, 117ff
   machine, 39ff
   performance, 38
protection (of materials), 84

qualitative formulation, 120
quantitative formulation, 120
query, 25ff, 73
   entry mode, 41
query-the-user, 75, 80, 83
quicksort, 51

range queries, 27
rapid prototyping, 75, 100
reasoning, 14, 21
   deep, 57ff
   shallow, 57ff
recognition (of lines), 94
recursive
   definition, 29
   query, 29
relation, 25, 27
   input/output, 121

re-planning, 140
resolution, 74
retrieval, 25
rule, 64, 73, 103, 153
   syntax, 64, 68
rule-based representation, 71
rule-based system, 17ff
   limitation, 17ff

simulation, 118, 121
skeleton, 92
slot, 124, 149
Smart Data Interaction package, 109
specification, 78
   executable, 100
speedup, 53
stack, 45
structural dependency, 120
structural description, 119
structural representation, 58
subclass, 124
substation (frame), 150
superclass, 124
switchgear station, 149ff
symbolic processing, 12
synchronization, 45

taxonomy, 151
technical system, 147
technology transfer, 69ff
term (Prolog), 41, 107
testability, 122
test-point, 59
time constraints, 134
tools
   configuration, 70ff
   for expert systems, 69ff, 98ff, 111ff
troubleshooting, 20, 57ff
TWAICE, 111ff

unification, 37, 49, 74
   parallel, 48
   schedule, 49

value constraint, 152
variable, 73
vision (computer), 91ff
VM/Prolog, 57

world models, 133

DATE DUE